Forage Conservation and Feeding

FRANK RAYMOND
GORDON SHEPPERSON
RICHARD WALTHAM

Illustrations by
CHRISTOPHER RAYMOND

FARMING PRESS LIMITED
FENTON HOUSE, WHARFEDALE ROAD, IPSWICH, SUFFOLK

First Published 1972

ISBN 0 85236 022 3

This book is set in 10pt on 11pt Times and is printed in Great Britain on S.E.B. Antique Wove paper by The Leagrave Press Ltd., Luton and London.

CONTENTS

APPENDICES

ILLUSTRATIONS

PHOTOGRAPHS

DIAGRAMS

FOREWORD

by Dr. GORDON R. DICKSON
Principal,
Royal Agricultural College,
Cirencester, Glos.

FOR TOO long our ability to exploit the potential of grassland through conservation and winter feeding of livestock has lagged far behind expertise in the production of herbage and its use through the grazing animal. In this book the authors indicate the means by which this deficiency can be overcome, and in so doing, make a valuable contribution to ensuring the success of our grass-based livestock industry in the more competitive environment into which farming is moving with the advent of EEC membership. Not only is this essential reading for livestock producers, but also for those who seek to benefit from the ley break in arable systems by producing grass products. Furthermore, in reviewing current research and practical innovations, and in indicating likely future developments in conservation and feeding, the authors have compiled a text book of immense value to all who seek greater understanding of this fascinating subject.

GORDON R. DICKSON

July, 1972

PREFACE

BRITISH AGRICULTURE now faces its biggest challenge. Soon it will lose the protection of the Agriculture Acts, which were designed to ensure plentiful and cheap food for the consumer. At the same time they have given the farmer a stability previously unknown. In their place will come the Common Agricultural Policy, with its 'free' exchange of produce between members of a market which is protected, by levies, from imports of cheaper food. But within the Common Market there will still be competition, each member striving to exploit its own resources of soil, climate and skill; there is wide agreement that for the UK this could lead to a greater emphasis on feeding grass and forage crops for meat and milk production.

How well equipped is British grassland farming for such a change in emphasis? For summer grazing many of the improved techniques such as controlled grazing and planned fertiliser use are already in commercial practice, and the main need is for them to be more widely adopted. The real problem arises in winter feeding, because the quality of much of the hay and silage now fed is so poor that virtually the whole 'production' ration comes from cereals and compounds. This has seemed unimportant to many whilst concentrate feeds have been relatively cheap. But these feeds will be much more expensive within the EEC, compared with the expected returns from meat and milk, and it is this that has led to the present interest in the more effective use of conserved forages in winter feeding—and to the present book.

But why three authors? Very simply, because not one of us knew enough about the various aspects of growing, harvesting, storing and feeding forage crops to be able to write this book alone. And this has been confirmed by the new information each of us has gained from the contributions of the other two!

In writing this book we have drawn widely on the experience and experiments of many colleagues, especially at Hurley and at Silsoe, and on the practical advice of many friends in farming. In typing and preparing the manuscript we have been helped by many others, and in particular by our wives. To all of them our most sincere thanks are due. We hope the chapters that follow will make a useful contribution to the technical information that farmers will need as we prepare to enter Europe.

FRANK RAYMOND
GORDON SHEPPERSON
RICHARD WALTHAM

May, 1972

Chapter 1

THE ROLE OF CONSERVATION IN BRITISH FARMING

CONSERVATION is the main operation in agriculture, for the harvesting and storage of cereals and potatoes is just as much conservation as the making of hay or silage.

In most areas of the world seasons of vigorous plant growth alternate with seasons of cold or drought, when plants do not grow. So man has always stored grains or roots, grown in 'summer' to feed himself during the 'winter'. But the storage of winter feed for animals is much more recent.

In this country, until the seventeenth century the consumption of meat and dairy products during winter was practically confined to butter, cheese and salted meat stored from the summer. Some forage stored as hay and straw was fed to the oxen kept for ploughing and later to provide the principal, and tough, supply of beef. Most sheep and pigs outwintered on rough grazings and in forests, and very few productive cattle or dairy cows were fed in winter, and these only for the tables of the rich.

This situation changed very slowly. The summer grazing of both cattle and sheep was improved with the Enclosures, but the aim in winter was still to limit loss in liveweight and condition, rather than to get production.

A few farmers grew turnips, a few made better hay. It was not until the 1800s that the upsurge in demand for meat from the new industrial cities made it necessary to *feed* rather than to *maintain* animals in winter. However, this feeding was not based on conserved forages. For by now farmers had available large quantities of cheap grain from Europe and America, and oil-cakes from Russia and Africa. This meant that they were able to develop livestock systems

15

in which their animals grazed in summer but were fed on coarse fodders and 'concentrates' in winter.

HAY-AND-CAKE FEEDING

For a while restrictions slowed down the import of cheap grain, and there was some interest in making better hay and silage, but after the repeal of the Corn Laws in 1846 the winter feeding of cattle and dairy cows in the UK became almost completely based on hay and cake. Thus although the RASE organised a competition in 1867 to help the introduction from France of the new system of 'ensilage', the incentive to improve forage conservation had virtually disappeared.

Apart from short periods during the two World Wars, this situation has remained almost the same until the present day. Despite all the advice to make better hay and silage, the feed trade still sells to home farmers each year some 3½ million tons of dairy concentrate to help produce 2,700 million gallons of milk—just about 3½ lb of concentrates for every gallon produced, summer and winter. When we reckon that home-grown barley is also fed, this says little for the productive value of our hay and silage—or of our pastures.

SUBSIDISED CEREALS

So there grew up the convention that conserved forages are fed for 'maintenance', and concentrates for 'production', and this has been enshrined in most textbooks on animal feeding. Official policy since the last war also seems to have been based on the idea that the average livestock farmer is incapable of making better hay or silage, and as a result cereals and concentrates have been subsidised at prices at which they could profitably be fed for the whole production ration. In fact in the early 1960s they were cheap enough to make up part of the maintenance ration as well, as in the barley beef and dairy straw balancer systems.

Of course, the enthusiasts made better hay and silage—and could show lower winter feed costs per gallon. But in many cases they would have had less bother—and made more money—by increasing their summer stocking rates, conserving less grass, and buying in more subsidised concentrates. In effect, the low price of cereals, in relation to returns from meat or milk has distorted the whole pattern of grassland development in this country in the last 20 years.

Many of the improved methods of forage conservation described in this book are not new; thus the advantages of tedding grass immediately after it is cut for hay were shown years ago, yet vast acreages of hay are still left for days untouched in the swath. Very simply it just has not been necessary to adopt better methods in order to make a profit.

This situation is unlikely to continue. In the long term the increasing food demands of world populations will mean that less cereals and oil-seeds are available for animal feeding—and what are available will go to pigs and poultry, which can use them more efficiently than cattle and sheep.

EFFECT OF COMMON MARKET

But a more immediate change is likely to be forced on the British farmer by entry into the Common Market; after a long period during which he could not afford *not* to feed cereals, he has five years in which to change to a system in which he will be unable to afford to feed large amounts of cereals to his ruminant animals.

We know that cost comparisons are dangerous. But the best estimates are that net returns from meat and milk in the EEC will be about 20% above present returns in the UK, whereas feedstuff prices will be over 40% higher. Systems of animal production will have to change to remain profitable.

The theme of this book is that one of the main changes must be the improvement of our methods of conserving forages for winter feeding, both by the wider adoption of methods already in use by the enthusiasts and by the application of new information from research. Unfortunately research activity is comparatively recent because research on conservation, just as farm practice, has been badly inhibited by the general view that hay and silage are maintenance feeds. As a result both research and practice have started with the idea that practical conservation systems must be *cheap*.

This has had the serious consequence that there has been little research on conservation methods which might involve capital investment. We remember the uproar in 1965 when Bridget's Experimental Husbandry Farm started investigations with tower silos! But research should first be concerned with biological problems—in the case of conservation, with the factors that determine the losses in different processes and the nutritive value of the products. Only when these are understood can the economics of different processes be sensibly evaluated.

The reverse has been too often the case; the idea that forage conservation must be cheap has restricted research activity not only in this country, but all over the world. By cheap we mean knocking hay about in the field in an indifferent climate until it is fit to bale, and putting wet grass into a hole in the ground and calling the product 'silage'.

Such methods lead to the high losses which have come to be

accepted as almost inevitable in the conservation process; few farmers would accept losing 30-40% of their barley or potato crops, yet these are normal losses with hay and silage. With this level of loss the reaction is to cut a big mature crop (which is also fairly dry and easier to conserve). Mature crops are of low feeding value, and in the process of conservation much of the feed value they do contain is lost by leaf shatter, leaching by rain and moulding, so that the feed value of the conserved product is even lower. And obviously a feed of such low value can justify only a cheap method of making.

This is a vicious circle. Cheap methods lead to high losses and low value products; low value products demand cheap methods. Perhaps this has not mattered as long as alternative feedstuffs could be fed economically; we believe the whole picture needs re-thinking as we enter a period of higher feedstuff prices.

This is not to suggest that good conservation methods *must* be expensive; but we are unlikely to find better methods if our initial thinking about the problem is restricted by the idea of cheapness. In fact, as we show in later chapters, some of the recent developments involve only quite minor changes from present systems; but these changes have resulted from a better understanding of just what we are trying to do—in improving silage fermentation for instance —and much of this understanding has come from studies which *themselves* have not been concerned with either economics or practicability.

Some developments do of course involve investment—in tower silos and high-temperature grass-drying, for example. But again it is not possible to decide whether such investment is justified until the full biological possibilities of these systems have been studied. Thus recent research shows that the high potential of feeding systems which include dried grass may well justify expensive drying equipment in certain farming systems. This conclusion could not have been foreseen when research on dried grass was restarted in 1965.

To understand the role of these new systems some knowledge of the principles underlying them is useful, and the early chapters are concerned with these. This knowledge is not essential—it is possible to use a chemical additive to make better silage without knowing how the additive works. But the most effective operators will be those who understand not only what to do, but also why they are doing it. While some readers may prefer to read just the chapters on how to make and feed conserved forages, we hope that most will want also to read about how and why the methods described were developed, so that they can apply them more effectively.

Chapter 2

THE PRINCIPLES OF CONSERVATION

GRASS CONTINUES to 'live' for a while after it is cut. In fact in bright sunlight it may continue to photosynthesise and increase in dry weight for several hours. But soon the cut crop dies; the plant cells lose their rigidity, the sugars in the plant juice start to oxidise and the proteins to break down. At the same time bacteria and moulds, which are always present on the surface of the grass but can do little damage to the living plant, begin to attack and decompose the dying tissues.

Conservation aims to check these destructive processes rapidly and completely, and to preserve as much as possible of the yield and feeding value of the crop. Two main processes are used: the crop is either dried, by haymaking or dehydration, to a stage at which both chemical breakdown and microbial action cease; or it is preserved at a high moisture content by the action of acids or other chemicals in the process of ensilage. Some knowledge of the background to these different processes is of great help in choosing the most suitable method for each particular enterprise, and in carrying it out successfully under farm conditions.

Haymaking

Haymaking has been defined as a means of producing a 'stable product of adequate nutritive value with the minimum of loss and at a reasonable capital and labour cost'. It is a method of conservation in which from 70-95% of the water present at mowing is removed by sun and wind whilst the cut crop lies in swath or windrow in the field. Some excess water may remain in the crop as carried from the field; this must be removed before the hay will keep safely and without further serious loss of feeding value. Heating and moulding in store must be avoided, and this may be done by artificial ventilation in stack or barn.

19

Final safe storage moisture content (mc) will vary with crop quality. Leafy immature hay with a high sugar content will need to be reduced to about 12% mc, either by forced ventilation or by natural heating and convection; even then there is a real risk that it will pick up water from the atmosphere during prolonged storage. Mature, more fibrous, crops with less readily-available carbohydrate to cause heating, will store safely at 15-18% mc and are less hygroscopic.

Grass cut for high-quality hay has a moisture content of about 80%, and a total of 3·25 tons of water must be removed to produce one ton of hay at 15% mc, the level to which it normally falls during the 6-month storage period before feeding. Allowing the grass to grow to a more mature stage increases yield and reduces the feeding value of the dry matter; but it also reduces its mc to about 75% and the amount of water which has to be removed to 2·4 tons per ton of dry hay.

The normal aim is to bale hay from the swath at not more than 25% mc, at which level less than 3 cwt of water per ton of dry hay remains to be lost from bales stacked in the field or in store. Hay removed from the swath at higher levels of mc, in an attempt to reduce nutrient losses, requires treatment either to dry it further, or to enable it to be stored damp. The amount of excess water increases rapidly as mc increases. So at 35% mc 6 cwt of water must be lost per ton of hay to be stored at 15%, but this figure increases to 8·3 cwt at 40%, 10·9 cwt at 45% and 14 cwt at 50% mc.

ENSURING A GOOD HAY CROP

For a good hay crop therefore, from 6 to 8 tons of water per acre have to be evaporated. Three-quarters of this can be removed on the day that the grass is mown, given suitable treatment and weather. Drying rate from time of cutting down to about 65% mc is almost constant, as water is lost easily from the surface and the outer cells of the plant; but in practice high drying rates can be obtained only if air can penetrate the swath and remove this water, and also the water which is formed from the breakdown—by oxidation—of plant 'sugars'.

In this early stage a tightly-packed swath can rapidly become much *wetter* than it was when first cut. Thus immediate action must be taken, by tedding for example, to help speed up the loss of water. In a swath the moisture content can differ between the top which is exposed to drying air and the bottom which is in contact with the wet soil; prolonging this phase of drying substantially increases respirational losses from the still-wet crop.

Mechanical damage and consequent loss caused by treatment to the freshly-mown crop is low because stems, petioles and leaves are tough, resilient and not liable to shatter.

As the moisture content falls from 60 to 30%, water diffuses rather slowly from the thicker stemmy parts of untreated plants; but whilst stems remain at a high mc leaves dry quickly, leading to a wide variation in mc between different parts of both the swath and the individual plants. Leaves become brittle and may break away from the stem and be lost, especially if severe mechanical treatment is applied below 40% mc when variation in mc in the swath is very high.

BENEFITS OF MECHANICAL CONDITIONING

The early application of a mechanical conditioning treatment which moves the swath as a whole, but which also crushes or lacerates the stems and thicker parts of the plant without too much effect on the leaves, does much to improve the evenness of drying at this stage. It has an even greater effect as mc falls below 30%, when the removal of the remaining water is difficult and is largely influenced by current weather conditions, particularly relative humidity. Often the application of these mechanical conditioning treatments causes hay to dry to a lower level of mc than could be obtained in a crop which has only been tedded or turned.

In practice, drying to less than 20% mc in the swath is very difficult, and even when possible it is usually associated with high leaf loss and a consequent disproportionate loss of protein. In any event rain falling on nearly dry hay, whether or not it has been tedded or conditioned, causes severe loss of soluble nutrients by leaching and reduces its overall feeding value, in particular its energy content. Prolonged wetting at this level can increase mc in the crop to as high as 70%, regardless of type of swath and treatment, and this can lead to moulding and rotting. Even if such hay eventually dries out it may have a feeding value no better than straw.

OVERHEATING PROBLEMS

Total loss of dry matter during the field-making process is often 30% or more, but even when hay still has a relatively high feeding value at the time of baling, this value can be reduced by respiration and other losses which occur in field heaps and in store. Although the rate is reduced as the temperature rises above 90°F (32°C) respiration continues until plant cells begin to die at about 113°F (45°C). Above this level heating continues owing to the activity of bacteria and moulds; this can cause additional nutrient losses of up to 40% of the total available.

Heating to more than about 160°F (70°C) signals the start of chemical oxidation, after which the admission of air to the stack causes a rapid temperature rise which may be followed by spontaneous combustion and total loss. Happily this stage is seldom reached, but the problem of brown overheated hay, or mouldy hay which can create a health hazard and has low digestibility and energy values, is common; although accurate figures cannot be obtained it has been estimated that up to 80% of the hay made in wetter areas in many years falls into this category.

It is therefore desirable, if quality is to be retained, to remove hay from the swath at the earliest possible stage and to evaporate the remaining moisture by some method of assisted drying. Table 1* indicates practicable levels at which this can be done, provided that effective treatment is given to ensure that the hay does not heat, mould and deteriorate in store.

Where hay has to be collected from the swath at above 30% mc it is more important than ever to ensure that field drying takes place evenly, because failure to do this creates difficulties in subsequent drying. There is also now much interest in the use of chemical additives to allow hay to be stored at higher moisture contents, and again uniform field drying is essential for their effective use.

ROLE OF HAY ADDITIVES

Efficient additives can prevent the development of moulding in damp hay (30-50% mc), and also have the important role of stopping respiration which wastes sugars left in the hay. It also produces water, which increases crop moisture content, and heat, which encourages mould development. At the least, moulding spoils palatability of the hay, but there is a much more serious aspect. If the hay is held between 122°F (50°C) and 158°F (70°C) for a long period, potentially harmful fungi develop. These produce toxic metabolites and spores which, when inhaled, cause Farmer's Lung in man and ulceration and mycotic abortion in cattle.

A number of hay additives which might check mould development, without having any adverse effect on feeding value, have been tested; the most promising has been propionic acid, but limited success has also been achieved with some mixed additives. Laboratory results have been encouraging and have indicated that the application of about 1% by weight of propionic acid can control moulding at hay moisture contents from 30-50%; it seems likely that a 2% application would effect control in baled hay.

* *All Tables referred to throughout the text are contained in Appendix 3.*

Unfortunately there is very little lateral movement of these chemicals in wet hay, so that effective control in practice requires a method of spreading the additive uniformly. An expensive alternative would be to increase the application rate; spraying with a dilute solution of the additive might also be considered. But the urgent requirement is for an efficient means of application using the minimum possible amount of additive; it seems likely that either a commercial or an independent research and development team will produce suitable equipment within the next few years. Until this is available hay additives will play only a limited role in improving haymaking.

Various chemicals, in both liquid and powder form, have been used on experimental and commercial farms with widely varying results, attributable partly to the effectiveness of the chemicals and partly to problems of distribution. It is likely that any product containing a fungistat will have some beneficial effect, but apparent advantages reported with some chemicals have almost certainly arisen from a simultaneous improvement in the mechanical techniques of haymaking employed, rather than from a direct effect of the additive.

MOST IMPORTANT PRINCIPLE

Perhaps the most important principle of all in haymaking is to try and match the method used to the expected drying rate and to the total amount of drying which is likely to be practicable in the swath. These factors will in turn depend very much on the length and the frequency of dry spells, and the mean daily relative humidity in any particular area. A study of past weather conditions, however, although historically of interest, confirms what the farmer already knows, that even in the more favourable lowland areas of Britain there is only one year in four when conditions allow a wide choice of mowing time coupled with a good chance of producing high quality hay, unless one of the newer improved techniques is used.

Advice on the approach of fine spells, coupled with local weather knowledge can, however, be very useful where methods of haymaking which lead to reduced swath exposure time are chosen—even though this can mean that to exploit a fine spell to the full, mowing may have to be done whilst rain is still falling!

High temperature dehydration

The use of heated air to evaporate most of the water present in the freshly mown crop substitutes high capital investment and the expenditure of energy from oil, gas or electricity for the losses incurred during the natural drying of forage for hay production.

Respiration losses are almost completely prevented by drying soon after cutting. Other losses in the process, which may be caused by burning and scorching of the leafy material in the drier and by the escape of fine dust in the processing and handling stage following drying, vary from about 3% in a well-designed drier up to a maximum of 8%.

High temperature drying goes much further than this, however, because it enables a younger high-digestibility crop of high moisture content to be dried at times when the removal of water in the field would not be possible; as the total drying costs involved are high it is essential that only crops of at least the very best hay or silage quality should be set aside for conservation in this way. In practice much of the herbage processed through high temperature driers could only be made into hay or silage with the greatest difficulty because of its exceptionally high mc, frequently over 85% and sometimes up to 90%.

The water which must be removed to reduce the crop to a safe processing and storage level is known as the 'drying load', and can be conveniently expressed in terms of tons of water per ton of crop dried to a particular level of mc—usually about 10% with a variation of $\pm 2\%$. It is normal to rate the output of driers on the assumption that they will dry from a moisture content of 80% down to 10%, and at this rating $3\frac{1}{2}$ tons of water must be removed from $4\frac{1}{2}$ tons of wet crop to produce each ton of dried material. Mean mc over the season is commonly nearer 82% and this small increase in wet-basis mc increases the seasonal drying load by about 14% above the standard rating, up to 4 tons of water per ton of dried crop. Wet grass of 85% mc has a drying load of 5 to 1 and at moisture contents above 88%, when drying load is 6·5 to 1, it is doubtful if drying can be economically justified.

WILTING REDUCES DRYING LOAD

An alternative to incurring the low rate of output and the high costs resulting from excessive drying loads is to mow the crop and remove some of the water by wilting, either in the field or by the application of some form of mechanical extraction technique. Reducing mc to 75% for example reduces drying load to 2·6 to 1, and at 70% it is only 2 to 1. There are technical and management problems associated with wilting, depending on the type of drier used, but these will be more appropriately discussed in the chapter on grass drying (Chapter 8) together with the effect of variation in drying load on performance and drying costs.

Suffice it to say here that whenever wilting is practised continuously it will always show a net reduction in drying load over the whole season, although when it is most needed it is often rendered ineffective by the very weather conditions that have caused the high moisture content in the crop. Gains in terms of drying load must be offset against any respiration and mechanical loss of crop during the swath wilting: these losses are often between 5 and 10%.

If mechanical extraction of water by the application of pressure is used, from 10-15% of the most valuable part of the dry matter is likely to be extruded in the juice. Some attempt must be made to utilise this juice, as otherwise the loss will more than offset any economies made in terms of fuel consumption and in the lower capital and labour cost arising from increased throughput.

As grass matures its moisture content falls, partly because of the physiological make-up of the individual plant and partly because on longer days during the height of the summer it will usually carry much less surface water. However, the rate of drying of more mature crops is likely to be slower than that of very young herbage, because leaves dry more rapidly than stems, and drier inlet temperature may then have to be reduced to avoid scorching. Hence the cost of water removal may not be decreased as much as would be expected from a theoretical consideration.

IMPROVEMENTS

Nevertheless, the benefits from a controlled reduction in moisture content before high-temperature drying are so evident that considerable effort is justified in the study of improved methods. One possibility is the use of a combination of field wilting with mechanical pressing, the latter, with any consequent loss of nutrients in the expressed juice, being applied only when wilting has not been effective. There is also the possibility that the 'shredded' crop likely to result from pressing will flow more uniformly through the drier than the usual chopped crop, in which leaves which dry rapidly may be attached to stems that dry more slowly. Another area of study is pre-treatment of the cut crop before pressing, so as to reduce the amount of nutrients expressed in the juice.

A further advantage of a system of field wilting coupled with a 'back-up' mechanical pressing stage would be that this would allow all the crop to be dried from a relatively narrow range of mc, compared with the wide range—70-88% or more—encountered when freshly-cut crops are dried. As a result drier design and operation might be simplified, and output balanced more nearly to the capacity of cobbing or pelleting equipment. Further, the planning of cropping and harvesting programmes should be improved, with a resulting reduction in day-to-day variations in drier throughput.

Ensilage

Quite distinct from these drying methods is the process of ensilage, in which the wet crop is preserved by chemical action. As the Ministry Bulletin No. 37 notes, 'it is not necessary to be a bio-chemist to make good silage, but some understanding of the silage process is certainly helpful in developing a successful silage system on the farm'.

A great amount of research and development has been carried out on silage since it was first introduced into Britain a hundred years ago. Most of this has been concerned with the preservation process and with the reduction of losses, and only recently has full recognition been given to the importance of the feeding value of the product, and the effect that different methods of ensilage can have on feeding value. Because of this we find that some earlier methods, which appeared to give efficient preservation, cannot now be recommended because of the limited usefulness of the product as an animal feed.

FIRST ESSENTIAL

As cut grass decomposes it heats up, and this heating in turn makes the decomposition proceed further. This is largely because, as warm air begins to rise from the heap of heating grass, it draws in fresh cold air, which further speeds up the heating process—rather like a draught through the fire. Thus the first essential is to stop air passing through the heap of grass. This can be done both by consolidation, which compacts the grass firmly so that movement of air is prevented, and by covering the heap with a plastic sheet whenever filling is not in progress.

Some consolidation generally occurs while the cut crop is being loaded into the silo, but undoubtedly the most effective method of controlling heating is to cover the surface of the heap; this acts, not by preventing fresh air getting into the crop, but by preventing warm air escaping, and this principle is an important feature in the Dorset Wedge sealed silo system, described in Chapter 7.

The first job of sealing then is to prevent air moving through the cut grass and causing heating; the second is to produce the oxygen-free situation within the heap of grass which is essential for effective preservation. But various moulds and bacteria can still continue to decompose the grass even in the absence of air; their activity must be stopped as rapidly as possible, either by making the crop acid, or by sterilising it.

PRESERVATION BY ACID

The most obvious way of acidifying the cut crop is by adding acid to it. The main process which has been used is that developed by

A. I. Virtanen in Finland, in which a mixture of sulphuric and hydrochloric acids (AIV acid) is added to the crop before it is put into the silo; other acids, such as phosphoric acid, have also been used. This method was at one time fairly widely used in Scandinavia, but it suffers from the very evident disadvantage that the operator must handle corrosive acids, and the perhaps less obvious disadvantage that the intake of the resulting silage is low.

Most silage-making, however, depends on the fact that cut grass normally carries a population of bacteria (the lactobacilli) which, in the absence of oxygen, can ferment sugars to produce lactic acid. These bacteria are relatively insensitive to this acid, whereas other bacteria and moulds—the putrefying organisms which decompose the protein in the grass—are inactivated as the grass becomes acid. This acidity is measured as pH, an index widely used by chemists, but confusing to the silage maker because *a decrease in pH represents an increase in acidity*. Also, pH is on a logarithmic scale: the decrease from pH 6·8 (the very slightly acid reaction of fresh grass) to pH 5·8 requires only 1/100th of the acid needed for a decrease to pH 3·8. Thus the initial fall in pH needs only a small amount of acid and can be quite rapid, but it slows down as more and more acid is needed to cause each further 0·1 unit reduction in pH—and also because the lactobacilli become less active at the lower pH values.

pH measures the concentration of acid in the water present in the silage. Unfortunately, the wetter the crop the lower the pH needed to ensure a stable silage, that is, the greater the amount of acid the lactobacilli must produce; for it is wet crops—leafy grass well fertilised with nitrogen, grass/clover mixtures, autumn grass etc— that are the most likely to have low contents of the sugars which the bacteria need to ferment to acid.

MEASURES TO PRODUCE GOOD SILAGE

Clearly, every effort must be made to establish in the silo the anaerobic conditions needed by these bacteria. But even when this is achieved, wet crops may not contain enough fermentable sugar to lower the pH to the required level for stable preservation. The activity of moulds and putrefying bacteria is then only slowed down; they continue to decompose protein, but more seriously they begin to decompose the lactic acid needed to preserve the silage. The pH then begins to rise as this acid is decomposed, and the damaging process of secondary fermentation sets in, producing the evil-smelling silage disliked by both animals and farm workers (and of course their wives).

For such crops special measures, in addition to efficient sealing, must be taken if good silage is to be produced:

(a) Wilting

Problem crops will not ferment properly because they are too wet and because they contain too little sugar. The first step must be, whenever possible, to reduce the moisture content of the crop by wilting in the field before it is brought to the silo: for less sugar has to be available for fermentation to ensile a crop of 25% dry matter content (stable at pH, say, 4·4) than a crop of 18% dm, which has to be acidified to a pH below 4·0 before it will store safely.

The process of field wilting is discussed in detail in Chapter 5. It has further advantages in that it reduces the loss of effluent from the silo (of great importance in avoiding pollution of rivers, etc), it reduces the weight of cut crop to be carted from the field and loaded into the silo, and it may increase the feeding value of the silage, because stock can eat more of wilted than of unwilted silage.

Wilting is of course essential when a crop is to be stored in a tower silo. Crop dry-matter levels in excess of 30% are generally advised, so as to reduce the pressure on the walls and the effluent loss, resulting from the weight of crop within the tower.

Because of the low moisture content, only a moderate degree of acid fermentation occurs before the silage reaches a stable pH, often in the range 4·5-5·0, which allows efficient preservation because of the complete absence of air in the tower. However, even when crops are to be stored in clamp or bunker silos, some degree of wilting before storage should be carried out whenever it is practicable.

But it may not always be practicable. Many smaller farms are equipped only with direct-cut harvesters; wet crops are often wet precisely because the weather is unfavourable for wilting; and the rate of wilting early and late in the season is slow, even when the weather is fine. Thus, to back up this desirable process of wilting, the silage-maker needs other aids, in particular the use of chemical additives which will make the ensilage process less dependent on the sugar and moisture contents of the crop.

(b) Addition of Molasses

If ensilage is uncertain because of a shortage of sugar in the crop, then it seems sensible to add extra sugar. The addition of molasses, at about 20 lb per ton of fresh crop (more if the crop is very wet or clovery) has been practised for many years. But success has been variable, mainly because of the difficulty of mixing molasses uniformly with the cut crop. Non-uniform mixing results in well-preserved silage interspersed with patches of poorly-preserved (high pH) material; and before the silage is fed the putrefaction from these patches can spread, so that much of the remaining silage is ruined.

(c) Chemical Additives

As ensilage involves chemical changes, it should be possible to improve the process by the addition of chemicals, and this possibility has been studied by many chemists for many years. The AIV process has already been noted, but most interest has been in reinforcing the natural fermentation process, rather than in replacing it by the use of strong acids.

Despite an immense research effort, chemical additives had virtually no effect on practical silage making until about 1965, and for precisely the same reason that molasses had failed—that no effective method had been worked out for mixing the chemicals uniformly with the cut crop. For most of the research had been done with mini-silos in the laboratory, far removed from the practical farm problems of making silage. Many will remember the enthusiasm with which sodium metabisulphite was introduced early in the 1960s —this new wonder chemical from the USA!—and its failure in practice, because it was spread by hand as the layers of grass were loaded into the silo.

The situation has been transformed by the development of effective applicators, working on the principle of feeding the additive directly into the cutting mechanism of the harvester (p. 112), but more practicable than with molasses because of the much lower amounts of additive needed per ton of fresh crop.

TWO SILAGE ADDITIVES

The introduction of efficient applicators has been rapidly followed by the introduction and marketing of new silage additives—some operating on better-established principles than others. A detailed review of these different additives would be quite out-of-place here: two are selected for discussion because they appear to be both effective and economical.

(1) Formic Acid

Formic acid as a silage additive was first studied in Germany in 1923, but it was hardly used in practical silage making until Naerland in Norway developed the simple applicator similar to that shown in photo 1. This was in 1965: by 1971 it was estimated that 95% of the silage made in Norway included formic acid. It was introduced into the UK in 1967, and was tested in detailed trials on experimental husbandry farms, at research stations, and on selected livestock farms before it was released commercially in 1969.

This work confirmed Norwegian experience that an average

optimum application rate was 6 lb of liquid (80 % formic acid) per ton of fresh crop, increasing to 8 or 10 lb with very wet or clovery crops. Research showed that this addition immediately reduces the pH of the cut crop to below 5·0, not low enough for safe storage, but low enough for a lactic-acid fermentation to take over and reduce the pH to a safe level, *as long as air is kept out of the silo*.

In many cases the final pH is only slightly lower than with a normal fermentation without additive; but because of the rapid early acidification there is little of the protein breakdown which occurs in the early stages of fermentation with normal silage. As a result the preservation remains stable over a long storage period, whereas secondary fermentation and decomposition may slowly set in with silage, apparently well-preserved, but made without the additive (fig. 1).

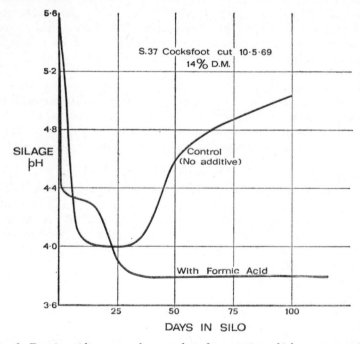

Fig. 1. Formic acid prevents the secondary fermentation which can occur when 'problem' crops are ensiled.

Clearly an additive such as formic acid is of most value in the ensilage of problem crops. But even with crops (e.g. wilted crops) which apparently ensile effectively without an additive, its use, possibly at a lower rate, is still advisable to give assurance of better preservation. And, as noted in Chaper 3, formic acid is likely to improve the feeding potential of many types of silage, by increasing the amount of dry matter that stock will eat, compared with silage made from the same crop, well-preserved, but without the additive.

The practical success of formic acid has already led to a renewed interest in silage additives, and much research, both government and commercial, is now under-way in the attempt to find even better additives. The farmer in the 1970s will be presented with many new products, and can reasonably expect reliable advice on their effectiveness—for previous experience suggests that some will be more effective and economical in use than others.

(2) *Additives based on Formaldehyde.*

There is already one interesting development. It has been noted that stock are able to eat more dry matter in the form of wilted than of unwilted silage. Some reasons for this are considered in the following chapter; but briefly the intake of high-moisture silage is low, either because of the extensive protein breakdown that occurs in badly-fermented (high pH) silage, or because of the very large amounts of acid that are contained in well-fermented (low pH) silage. When formic acid is used protein breakdown is largely prevented, because the pH of the crop is rapidly reduced below the level at which the undesirable bacteria and moulds can operate; as already noted, the intake of this silage may be higher than that of silage made without additive, but it still remains lower than that of the fresh crop, because the silage (low pH) contains a lot of acid.

This has led to the examination of chemical additives which will preserve the crop at a high pH, but at the same time prevent protein breakdown. Formaldehyde is such a chemical. Work by Wilkins and Wilson at Hurley has shown that the addition of 2 gallons of formalin (40% formaldehyde solution) to each ton of wet crop (18% DM) will give efficient preservation at a pH of about 5·5. The formaldehyde, a sterilising agent, rapidly kills most of the bacteria and moulds present in the crop; at the same time it combines with the protein and prevents its decomposition, which can occur even in the absence of bacteria. But there is little *positive* preservation of the silage, so that this additive is only likely to be effective if a strict system of sealing the silo against entry of air is used (Chapter 7). Studies have also been carried out with mixtures of formaldehyde and formic acid; while these give a slightly more acid silage, they could have the practical advantage of reducing the risk of secondary moulding at the silage face when the silo is opened.

Laboratory experiments with these additives were carried out in the winter 1969/70. Small-scale silos were made for feeding in the following winter, and the results were so encouraging that in 1971 larger experimental silos were made at Hurley, at two EHF's, and also by several adventurous farmers. Further work is still needed before the additives can be advised for general use, but first results do indicate the exciting possibilities they may offer for improved conservation of silage under practical farming conditions.

*   *   *   *

To date a high nutritive value, and especially a high intake level, in hay and silage have seldom appeared necessary because cheap, cereal-based supplements have been readily available. As cereal prices rise inside the EEC this will no longer be the case; and many have concluded that, because the farmer finds it difficult to make better conserved forages, milk and meat prices must rise sharply to cover the higher feed costs. We believe this need not happen if farmers in the UK can rapidly adopt the improved systems of conservation described here. And if meat and milk prices do rise sharply, because Community prices are high, we believe that livestock farming, based largely on conserved forages, will prove more profitable than that based on cereals.

Chapter 3

THE FEEDING VALUE OF CONSERVED FORAGES

HAY, SILAGE or dried grass are seldom used as the only feed for productive stock. But because, in the future, their unit costs on the farm are likely to be lower than those of most alternative feeds, the aim will be to include as much of them in the total ration as can be fed without reducing the level of animal production obtained.

This is not difficult with low-producing stock such as store cattle, rearing heifers, late-lactation and dry cows. It is with the high-yielding cow, the beef calf and the fattening steer that the problem arises of getting a high enough level of nutrient intake to satisfy production requirements.

Essentially, high nutrient intake requires a high intake of feed dry matter (intake) and that this feed is efficiently digested by the animal (digestibility), this being the main measure of the energy value of the feed. In addition, the feed must contain adequate levels of protein, vitamins and minerals. But under most practical feeding conditions it is an insufficient intake of digested dry-matter, or more accurately, digested organic matter (p. 34), by the stock being fed that limits their level of production. Thus if we are to increase the role of conserved forages in animal feeding, we need to understand the factors which determine (a) how much conserved forage animals eat and (b) how efficiently they digest this forage. Further, because conserved forages are likely to be fed as part of a mixed ration, it is also important to understand their possible interaction with the other feeds which may be included in the total ration.

Digestibility

The one equation in this book is included to describe feed digestibility:

$$\text{Digestibility } \% = \frac{\text{Food digested}}{\text{Food eaten}} \times 100 = \frac{\text{Food eaten—dung excreted}}{\text{Food eaten}} \times 100$$

that is, the lower the excretion of dung for each unit of food eaten the higher is the digestibility of that food.

Most experimental measurements of food digestibility have been made with sheep, by weighing the amounts of food eaten and of dung excreted over a 'balance' period, generally of 10 days. This however, is an expensive procedure, of little use to the adviser or the farmer who wishes to assess the digestibility of a batch of hay or silage. Consequently, much research has been directed at developing laboratory methods which will estimate digestibility by analysis of a small sample of the feed. The most precise of these is the *in vitro* digestibility method, in which the feed sample is digested with liquor taken from a sheep's rumen, to simulate the process of rumen digestion. But this is not well-adapted to routine use, and a simpler, though rather less precise, method is commonly used in advisory work—the analysis for modified acid detergent fibre (MAD fibre): the higher the content of fibre in the feed sample, the less efficiently will animals be able to digest that feed.

Many earlier values were quoted in terms of dry-matter digestibility, based on the dry weights of feed eaten and of dung produced. But the feed may also contain mineral materials which, though digested, are of no energy value to the animal, as well as soil and sand.

As digestibility is mainly a measure of the energy value of the feed, the decision was taken in 1967 to use 'digestible' *organic* matter, for it is only organic constituents of the feed which the animal can use for energy purposes. The measure adopted was D-value, defined as the percentage of digestible organic matter in the dry matter of the feed. From this it is possible, with reasonable accuracy, to calculate directly the Starch Equivalent (SE) value of the feed (the feeding unit most commonly used in the UK (p. 38), or the metabolisable energy, which has been considered as a possible alternative to SE. But digestibility (D-value) is much simpler to envisage 'in the field' than these two other measures, and is used in the following discussion.

THE DIGESTIBILITY OF FORAGES AT CUTTING

The digestibility of hay and silage can be estimated in the laboratory; but of far more practical use is the information now available on the digestibility of different crops while they are still growing in the field, which enables the farmer to predict pretty closely the digestibility of the hay or silage *he is going to make*.

It has long been known that, as the date of cutting of a given crop is delayed, so the feeding value of the resulting conserved crop is lowered. This is because the crop becomes less digestible as it becomes more mature.

Work at a number of centres, in particular at the NIAB at Cambridge, at the Welsh Plant Breeding Station, at the West of Scotland College of Agriculture and at Hurley, has now shown that the digestibility of growing crops can be *predicted* with considerable accuracy from a knowledge of the crop species and its stage of maturity at harvest. Maturity is often described in terms of the number of days from the date on which the crop reaches 50% ear emergence (in the case of grasses) and so can be applied to each particular species growing in the field and also to crops growing in different regions.

These relationships have also been extended to indicate the approximate levels of yield and of protein content of a wide range of forage species and varieties at different levels of D-value. With this information the farmer can decide the stage of maturity and thus the date at which to cut each crop, so as to obtain the best combination

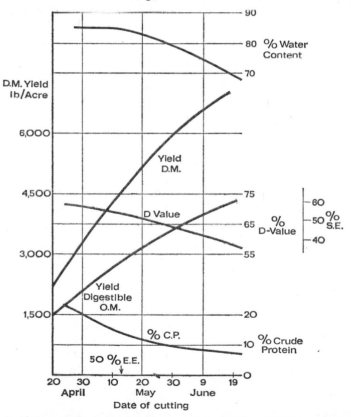

Fig. 2. Changes in the yield, digestibility and composition of S24 ryegrass during its first growth in the Spring. (Mean data at Hurley).

of yield and feeding value.

Results for a typical crop, S24 ryegrass, are shown in fig. 2. Up to mid-May (later in northern areas) digestibility falls only slowly, while yield is increasing at a fair rate. Then, at about 50% ear-emergence (that is, when ears have just emerged from half the fertile shoots in the crop) digestibility begins to fall more rapidly; but at the same time growth is at its most rapid. This indicates that, if a crop of very high digestibility is required, some sacrifice in yield must be accepted; if high yield is the main aim, then the crop will be of lower digestibility.

Clearly the decision made will depend on the class of stock to be fed, the other feeds available, and to some extent on the method of conservation to be used—young, highly-digestible grass is always likely to be more difficult to conserve than more mature, but somewhat less digestible, grass. But the important point is that fig. 2 can take some of the guesswork out of making this decision upon when to cut the crop.

Similar data, available for most of the forage crops grown commercially in the UK, are considered in Chapter 4. The most accurate information on digestibility is undoubtedly that for the first growth of the year. Much of the national hay and silage crop is made from such first growths; but much is also made from re-growths, either from an earlier grazing, or from hay or silage aftermaths. It is more difficult to predict the digestibility of these re-growths, because this depends to a large extent on exactly when the earlier cutting or grazing took place.

A re-growth of S24 which was grazed up to May 5th will tend to be stemmy, and its digestibility will fall rather rapidly; a re-growth from a silage cut taken on May 25th will contain relatively few seed-heads, and its digestibility will fall more slowly. But results recently published from Hurley[1], indicating the likely level of digestibility of re-growths of several grass species following first harvests taken on a range of dates, should help in deciding the dates at which these different re-growths should be cut to obtain forage of a particular digestibility.

Another practical problem arises when old pastures or older leys are to be cut, for their mixed botanical composition makes it more difficult to predict their digestibility. But we still know that, as they become more mature, such swards become less digestible; they often contain rather late-growing species, and a digestibility level similar to that of S23 ryegrass on the same date is probably a reasonable estimate.

[1] GRI Technical Bulletin No. 8, 1971.

THE DIGESTIBILITY OF CONSERVED FORAGE

The importance of this information on the digestibility of the growing crop is that it can be used to predict the digestibility of the conserved forage that will be made. In fact many experiments have shown that, with efficient methods, the digestibility of hay, silage and dried grass is very close indeed to that of the standing crop when it was cut.

But the proviso, 'with efficient methods', is vital. Work by Shepperson at NIAE in the late 1950s (Table 2) showed that barn-dried hay was rather less digestible than the cut crop, but that digestibility and protein content were both much lower with field-made hay. This was particularly marked under unfavourable drying conditions, because of the mechanical losses of leaf and the losses due to leaching by rain which occurred before the crop was fit to bale. This reduction in digestibility can be even more marked under farm conditions, because of heating and moulding in the stack when hay is baled before it is fit.

With silage a decrease in digestibility can result from effluent running from the silo, and from overheating. But the main damage occurs when field wilting is attempted in unsuitable weather, and a considerable reduction in nutritive value can occur when crops are wilted for high-dry-matter silage under unfavourable conditions.

Thus the digestibility of a conserved forage will nearly always be lower than that of the forage conserved, because of chemical changes and losses during the conservation process. The poorer the method, the greater will be the fall in digestibility, and the methods described in this book aim to reduce this fall to an absolute minimum. But from a knowledge of the method used and of whether the losses, e.g. in field wilting, were average or above average, a reasonable estimate of the digestibility of a conserved forage can be made, based on the estimated D-value of the cut forage, corrected as in Table 3.

A more accurate estimate can, of course, be obtained by sending a sample of conserved forage to an ADAS laboratory for analysis. But such analysis can be carried out on only a few of the tens of thousands of different lots of forage which are conserved each year. With most lots of hay and silage, estimates of D-value, based on information such as that in figs. 2 and 3 and Table 3, should allow these feeds to be used quite effectively in the winter ration.

DIGESTIBILITY VALUES AND STARCH EQUIVALENTS

In this chapter the feeding value of fresh and conserved forages has been examined in terms of digestibility, D-value, partly because it is D-value that has been measured experimentally, but also because the idea of 'digestibility' is much easier to grasp than any other measure of feed energy value.

However, ruminant feeding systems in this country are based on Starch Equivalents, and to make use of the information on forage D-values it is necessary to calculate SE values from them (although these values can only be approximate, as few actual measurements of forage starch equivalents have been made).

Table 4 gives the estimated SE values of feeds for a range of D-values, which are currently used by the nutrition chemists of ADAS. It can be seen that a decrease of about half in D-value leads to a decrease of over two-thirds in SE value, because animals use the digestible energy from feeds of low D-value less effectively than from feeds of high D-value; this is yet another argument for aiming for feeds of high digestibility. These SE values have also been included in figs. 2 and 3, and this allows an estimate to be made of the SE value of a cut forage in terms of its species and the stage of maturity (date) when it was cut.

Feed intake

Under many practical feeding conditions the contribution made by conserved forages to the rations of productive animals is limited by the low amount of these forages that animals are able to eat. As a result, large amounts of supplementary feeds may have to be given to bring total nutrient intake up to the required level. If forages are to make a bigger contribution to livestock rations, this problem of low intake *must* be overcome.

First it must be noted, that although some feeds are 'unpalatable'—that is, stock do not eat them because they dislike them—this is in practice seldom the main reason for low intake. The real problem arises with feed which animals quite obviously find palatable, but where the amount of feed they are able to eat is limited by some other factor.

THE INTAKE OF HAY

It has long been known by observant stockmen that animals are able to eat much more of a young, leafy hay than of an old, stemmy hay. The leafy hay is highly digestible, the stemmy hay is of low digestibility; and with a wide range of feeds we now know that, as feed digestibility decreases, the amount of feed that animals are able to eat also decreases. This means that *nutrient intake* changes very markedly with changes in feed digestibility, and lends added importance to the discussion in the previous section of the factors determining forage digestibility—and of the practical possibilities of predicting this from a knowledge of the species and stage of maturity of the crop being cut.

This relationship between intake and digestibility is most marked with fresh-cut (zero-grazed) crops, with crops conserved as hay, and with dehydrated forages. For example, young steers ate about 8·7 lb dry matter of well-made S24 ryegrass hay of 70% D-value, but only 7·5 lb of hay, made from the same species but cut 4 weeks later and of 60% D-value. The intake of digestible organic matter from the first hay was 6·1 lb $\left(8\cdot7 \times \dfrac{70}{100}\right)$, and from the mature hay 4·5 lb; in a practical feeding trial the first hay gave 2 lb liveweight gain a day, and the late-cut hay only 1·5 lb (Alder, GRI).

Why do cattle and sheep eat less of low-digestibility feeds? The reason is that the amount of feed that ruminants can eat is mainly determined by the capacity (volume) of the digestive tract, and particularly of the large first-stomach, the rumen. It seems that an animal will eat until a certain level of 'fill' is reached within the rumen. With highly-digestible feeds the level of fill is rapidly reduced, partly because these feeds are *rapidly* digested and broken down, and partly because there is little indigestible residue left when they *are* digested; as a result the rumen empties rapidly, and the animal can eat more food. In contrast, the rumen filled with feed of low digestibility empties very slowly, because the feed is slowly digested, and because large amounts of residue are left. Low-digestibility feeds occupy space in the stomach for a long time, and this limits the amount of such feeds that animals are able to eat.

DIFFERENCES IN INTAKE BETWEEN FORAGE SPECIES

For some time it was thought that the relationship between intake and digestibility was so precise that it should be possible to *predict* how much of a given forage an animal would eat merely by knowing the digestibility of that forage: in other words, nutrient intake could be 'read off' directly from graphs such as fig. 2. However, it is now known that there can be considerable differences in intake between different forages, even though they have the same level of digestibility. The most notable example is the higher intake of legumes (white clover, red clover, lucerne and sainfoin) than of grasses of the same digestibility. Against this must be set the generally rather lower digestibility of the legumes than of the grasses at the stage of growth at which they are commonly harvested—though an exception is white clover, which is normally *more* digestible than the grass with which it grows in the sward. But stockmen are well aware of the benefits from including some legume in hay mixtures—if only they could make the hay without moulding or losing all the clover leaf. The need for the improved hay-making methods described later was never more evident than with grass/legume mixtures.

Even among the grasses we also find differences in intake which could be large enough to be of practical importance. The most interesting example is the generally higher level of intake of feeds made from Italian ryegrass than from perennial ryegrass of the same digestibility. The most surprising example is the rather low intake of timothy—surprising because of the general opinion of the very high 'palatability' of timothy, particularly when it is made into hay. However, this observation could well be so. At the advanced stage of maturity at which most hay is still made, the digestibility of timothy hay may in fact be *higher* than of ryegrass hay, because timothy decreases in digestibility more slowly than ryegrass as these two grasses mature (fig. 3). This could outweigh any theoretically higher intake of ryegrass than of timothy at the *same* level of digestibility.

The reasons for the higher intake of some forage species than of others may appear of rather academic interest—but their understanding could well lead to the breeding of new varieties of even higher intake, and so make available to the farmer crops of improved potential for animal production. The major reason is in fact the same as that which relates intake to digestibility—the more rapidly a food is digested and leaves the rumen, the greater is the amount of that food that ruminants can eat.

We now know that when legumes and Italian ryegrass are eaten they are more rapidly digested than, for example, perennial ryegrass of the same level of digestibility. The differences in intake may appear small (although the intake of sainfoin is some 40% higher than that of timothy of the same digestibility) and they are certainly less important than harvesting at the right stage of digestibility, or than improving the overall system of conservation. But once these aspects have been mastered it may well be worth considering the advantages of growing forage species of inherently high intake potential, so as to increase the level of nutrient intake when they are fed.

THE INTAKE OF DRIED GRASS

A similar relationship, with intake increasing with forage digestibility, is also found when long dehydrated forages are fed. However, most dried forages are now packaged in some form before they are fed. Even if the forage is not hammer-milled, the process of packaging leads to a reduction in the particle-size of the dried forage; as a result the intake of the packaged forage is nearly always greater than that of the fresh crop, or of the long dried crop before packaging. This effect becomes more marked the lower the digestibility of the crop being dried, as does the effect of packaging on the production potential of the dried feed.

Table 5 shows the liveweight gains by cattle fed dried ryegrass cut on a range of dates. As the date of cutting for drying was delayed, the liveweight gains of the cattle fed the chopped dried grass were markedly reduced, because both digestibility and intake decreased with increasing maturity of the crop. At each date of cutting, part of the dried crop was also milled and pelleted before feeding. Voluntary intake increased in all cases, as did the resulting rates of liveweight gain. The greatest increase was found with the most mature feed, but it is important to note that, despite this, gains on the more mature pelleted dried grass were still much lower than on the less mature feed fed in the chopped form.

It now seems that, except in cases where the dried forage must supply the long 'fibre' needed to ensure good butter-fat levels in milk (p. 47), there is a real nutritional advantage in reducing the particle size in dried grass before it is fed. This can be achieved by hammer-milling before pelleting, or by compressing the chopped dried feed in a press in which grinding takes place during the process of extrusion ('cobs'). This nutritional advantage of small particle size has been found with beef cattle (Table 5) and with sheep fed on dried grass as sole feed; it also occurs when dried grass is fed as a supplement to a basal diet of hay or silage, the most likely way in which it will be fed in practice (p. 141).

THE INTAKE OF SILAGE

The biggest practical problem involving low level of feed intake is found with silage. There is now a great deal of evidence that ruminants are likely to eat much less dry matter as silage than as hay made from the same crop, and much less also than as the fresh crop from which the silage was made. Thus Stanley Culpin, at Drayton EHF, recorded cattle gains of only $1\frac{1}{4}$ lb/day when silage was fed *ad lib* compared with over $1\frac{1}{2}$ lb when hay made from the same crop was fed. The problem is most serious with silages of low dry-matter content, and the intake of wilted silage is generally higher than of unwilted silage made from the same crop.

Much research during the 1960s attempted to identify the causes of the low intake of silage; high water content, acidity, toxic constituents, butyric acid, etc were incriminated by various workers. But no single factor could explain all the observed cases of low intake. Then in 1970 Wilkins and his colleagues at Hurley showed that at least two factors were involved. With low-pH (well-preserved) high-moisture silage, intake is limited because of the large quantity of acid which the animal must take in with each unit of silage dry matter eaten; for some reason there is a limit to the amount of acid animals can eat.

At the other extreme, the intake of wet, high-pH, silage seems to be limited by breakdown products from the proteins decomposed by the putrefying bacteria characteristic of such silage; this breakdown is indicated by a high content of ammonia in the silage.

High-moisture silages of intermediate pH will also tend to have low intakes; as the pH rises the silage acids will have less effect on intake, but the effect of the protein-breakdown products will increase. Of the two, the latter effect seems to have the more serious effect on intake; and the revolting smell of putrefied high-pH silage is also surely unacceptable under modern farming conditions.

INCREASING SILAGE INTAKE

Although the mechanisms causing low intake are not fully understood, it does seem that to ensure high intake the silage-making process should aim to avoid both protein breakdown and excessive acid content. Yet high acid content (low pH) has always been the main target of the efficient ensilage process. This has led research workers to study methods of storing green crops at a relatively high pH, yet without the decomposition associated with protein breakdown. The most commonly-used method is to reduce the moisture content of the crop by wilting before it is ensiled; as noted on page 28 many crops of 25% dry-matter content and over produce a stable silage, under airtight conditions, with a pH above 4·5. This silage contains relatively little acid, and protein breakdown is largely avoided; as a result its intake by animals is higher than that of the corresponding unwilted silage.

A new approach has been opened up by the observation that the amount of silage that animals will eat is increased when formic acid is used as a preservative. As fig. 1 shows, the pH of this silage is if anything, lower than that of a control silage made without additive. But the formic acid also markedly reduces protein breakdown because of the rapid fall in pH during the early stages of the ensilage process and the prevention of secondary fermentation; this outweighs any effect on intake resulting from the higher acid content of the formic acid silage. This discovery has led to the search for additives which will give efficient preservation without the need for low pH, and yet which prevent protein breakdown; essentially the aim is to sterilise the crop rather than to preserve it by fermentation.

This work is only in its early stages. Studies at Hurley have already shown that formic acid applied at high levels (1½-2 gallons per ton), formaldehyde (40% formalin) at 2 gallons per ton, and mixtures of formic acid and formaldehyde will preserve wet crops of 20% dry matter or lower very effectively. And, as predicted, the intakes of the 'silages' made with these preservatives are higher than that of the control silages.

Promising results with these silages have already been obtained on a farm scale; but more extensive feeding trials are needed to evaluate the new additives, and to ensure that they do not lead to animal health problems, before their general use can be advised. Similar 'silages' have in fact been fed to dairy cows in Finland since 1969, without any problems arising, although the Finnish scientists have perhaps not exploited the potential for high intake indicated in the recent UK studies.

This new concept of preservation of wet green crops opens up new possibilities of improved conservation under practical farming conditions. Yet formaldehyde, like formic acid, was studied many years ago as a silage additive: but also, like formic acid, its practical use has only become possible with the development of efficient applicators and effective methods of sealing silos—that is with the combination of science with practice.

Conserved forages fed in mixed rations

These studies of some of the factors determining the feeding value of hay, dried grass and silage are an essential step in developing improved feeding systems. But only a step; for in practice these conserved forages will mostly be fed in mixed rations, and it is therefore important to know how far their 'basic' feed values may be modified by the other feeds used in these rations.

BALANCED RATIONS

The feeding of farm stock is based on the principle that (a) the nutrient requirements of an animal, in terms of energy, protein, minerals etc, can be calculated from tables of requirements for maintenance and production for the particular class of animal, and its level of production of milk and liveweight gain, and (b) these requirements are met by the nutrient content of the total ration, calculated from tables of the nutrient contents and amounts of the different feeds eaten. In this way deficiencies of particular nutrients, especially in the 'roughage' part of the ration, can be made up by cereals and concentrates rich in these nutrients.

This system worked well with rations based mainly on concentrate feeds, but problems arose when greater reliance was placed on home-grown forages during the 1939-45 war, and adjustments had to be made to the values for these feeds in the feeding tables. Part of the reason for this is that feed values are not truly *additive*, and it is important to understand some of the interactions between feeds if more effective use is to be made of conserved forages.

CONSERVED FORAGES AND CEREALS

It has long been known that cereals can reduce the digestibility (energy value) of forages with which they are fed. Recent studies suggest that this is because cereals tend to make the contents of the rumen rather acid; under these conditions the micro-organisms in the rumen which digest fibre become less active, so that the extent to which they digest the fibre in the forage part of the ration is reduced. But probably more important than the reduction in level of digestibility is the effect on intake; for at the same time there is a reduced *rate* of fibre digestion, and as a result the rate of passage of feed decreases, and the animal eats less forage. Because of this, cereals, fed as a supplement to forage, in practice partly *replace* the forage, so that the effective incremental value of the supplement is lower than expected. The most marked effect observed has been a decrease of 0·9 lb DM in the intake of high-digestibility Italian ryegrass hay for each extra 1 lb of rolled barley fed as a supplement, though the extent of replacement is generally less than this.

The level of intake of high-dry-matter silage is also reduced when cereal supplements are fed, and a similar effect may be expected with the new types of high-intake silage, described in the previous section. (Interestingly, supplements have much less effect on the intake of normal low-dry-matter silage, almost certainly because the intake of this silage is not controlled by bulk factors, and so is less sensitive to rate of fibre digestion).

This decrease in forage intake is perhaps unimportant when the quantity of hay or silage fed is restricted; it is more serious when high quality forages are being fed *ad lib*, with the aim of exploiting their high intake potential. Ways of minimising this effect of cereal-based supplements need further study; thus, for example, more frequent feeding of the supplement might help to reduce the sudden fall in rumen pH observed when concentrates are fed at milking time to dairy cows.

But the most promising approach is in the development of supplements which cause less decrease in forage intake when they are fed. And here some recent experience with dehydrated forages is of interest. These feeds do not reduce rumen pH as much as do cereals: and thus cattle and sheep eat more hay and silage when dried grass is fed than when an equal weight of barley is fed. This effect is most marked when the dried grass contains small particles; thus more silage is eaten with a supplement of pellets than with cobs or wafers.

In many cases the total dry-matter intake of rations comprising half silage and half pellets (on a DM basis) is as high as when the pellets are fed alone; the resulting levels of animal production, noted in Chapter 9, indicate the possibility of considerable reductions in feed costs with this feed combination.

However, to be realistic, the quantity of dried grass likely to be available in the foreseeable future will be enough to provide half the feed intake of only a few animals. The practical aim must be to feed limited amounts of dried grass to as many animals as possible; for this it is most likely to be combined with barley, in the form of a compound to replace part of the conventional compound which is fed as a supplement with hay or silage.

For quite unexpectedly, and unlike its effect with hay and silage, barley appears to *increase* the feeding value of milled dried grass. The estimated SE of dried grass is generally in the range 45-50, while the SE of barley is around 72. But quite small additions of barley appear to lift the effective SE value of dried grass markedly, as is seen in the results from feeding work at Drayton EHF, reported on page 145. A similar result has appeared from the dairy cow feeding experiments at EHF's, summarised by Strickland (p. 142). A compound, in the form of cobs containing equal parts of dried grass and rolled barley, with vitamins and minerals, has now been adopted by MLC as the standard ration fed to bulls at their performance testing centres (p. 146).

The reasons for this unexpected interaction between dried grass and barley are not yet understood. But its implications, considered more fully in later chapters, are considerable: thus the dried-grass/ barley compound used by MLC is being marketed at an appreciably lower price per ton than conventional concentrates. But it must be emphasised that the dried grass used in all these combinations with cereals has itself been of reasonably high feed value, i.e. a D-value of 60 or over. Where dried grass of lower digestibility is used, the combination with barley is likely to be less productive. Further, dried-grass/barley compounds may not replace all the high-energy compounds needed in early lactation to stimulate high milk yields by productive cows.

The protein value of conserved forages

In general it seems that the most effective method of using dried grass will be in combination with cereals (either mixed together at the time of feeding, or included in a compound pellet) fed as a supplement to hay or silage. One of the major advantages of this combination is that maximum use can be made of the high crude-protein content characteristic of dried grass, for, even with the improved methods described in Chaper 6, the protein content of hay is likely to be in the 10-12% range. Silage, cut somewhat earlier, will generally have a higher protein content (12-15%), but as a result of the normal fermentation process, some of this protein will be in a highly-soluble form which is wastefully used by ruminants*. Most lots of barley are below 12% crude protein, with the majority below 10%. Because of this, rations of hay and silage with barley must be supplemented with protein if they are to be fed to high-yielding dairy cows and young stock, which need more than 12% protein in their diets. The most commonly-used protein supplements are the oil-seed meals and fish-meal. But these must be imported, and are also expensive; we believe they can now largely be replaced by the protein in dried grass. In fact recent research indicates that the heating of the protein in grass which occurs during the dehydration process improves its value for ruminants. This may be one of the reaons why dried grass combines so well with silage and with barley, both of which are deficient in protein.

Another aspect of protein nutrition which is of increasing interest is the possible use of non-protein nitrogen as a substitute for part of the protein in ruminant rations. The most commonly used form is urea. When this is eaten it is broken down to ammonia by the rumen bacteria, and if the rest of the food contains a good supply of available energy these bacteria are able to use some of this ammonia for their own growth. The animal can then, in turn, use these bacteria as a source of protein. Because of this need for energy, urea is mainly fed in combination with cereals, and is included at a low level in many commercial compound feeds, and in mineral licks combined with molasses. It has been much less successful as a supplement to low-protein forages, because these are so often also of low energy content (that is, low digestibility); this can lead to the serious condition of urea toxicity, when ammonia is produced within the rumen more rapidly than the bacteria can utilise it.

This indicates that urea is most likely to be effectively used when it is fed with forages of low protein content but of high digestibility. The most promising of these is forage maize; recent work at Hurley has indicated that urea, uniformly mixed at a level of $1\frac{1}{2}$% of the dry weight with maize silage before it was fed, reduced the need for supplementary protein by young cattle (p. 144).

American work has also suggested that urea is effectively used when dehydrated lucerne is included in the ration. A similar benefit was indicated in the cattle experiment, where daily gains by the youngest cattle were increased to 2 lb when the silage was supplemented with both urea and lucerne pellets. Current research into the

* This may not be the case in silages preserved with formaldehyde.

way these feeds interact could lead to further improvements in the use of urea in ruminant rations, and these in turn to useful economies in feeding.

Fibre in ruminant rations

Ruminant animals are particularly adapted to digesting fibrous feeds, but must they have fibre in their diet? To answer this we need to distinguish between fibre as a chemical fraction and fibre as the structural component of 'fibrousness'. Too much chemical fibre reduces the digestibility of the ration; but a certain amount of 'long' fibre is generally essential for the effective operation of the ruminant digestive system. This first became evident with 'barley beef'. Barley feeding without any roughage, which had worked in the strictly-controlled conditions of the Rowett Institute, proved a hazard on the farm because of losses due to acidosis and bloat when cattle did not ruminate. This was prevented by feeding a pound or so of hay or straw, which was enough to stimulate rumination.

Similarly, beef cattle can be fed for long periods on dried grass pellets, containing no long fibre; but it is much safer, and management is easier, if they are also fed hay or straw—but only in small amounts, so as not to limit the intake of high-digestibility dried grass. Thus bulls in the MLC performance tests are given 4-5 lb of hay daily, in addition to dried-grass/barley cobs, because the fibre particles in the cobs are not large enough to ensure proper rumen activity.

In the case of the dairy cow, however, there is a positive need for long fibre in the diet, to ensure the correct rumen fermentation to give good butterfat levels (fairly fibrous diets produce acetic acid in the rumen which is needed to synthesise fat). In most traditional dairy rations plenty of fibre has been supplied by the 'roughage' fed for maintenance. However in the mid-1960s interest arose in Europe (especially in Denmark) in the possible use of dried grass to replace conventional hay and silage. In that case it was concluded that the dried grass must provide the long fibre needed by the dairy cow, and several machines were developed to produce dried grass 'wafers'— fairly large packages which contain a proportion of long particles.

For several years this idea was foremost in discussions on dried grass, but it is doubtful if it is now generally applicable. Firstly, feeding experiments have shown that there is seldom enough long fibre in 'wafers' to ensure adequate butterfat levels: chopped straw has to be fed as well, and it seems preferable to feed more chopped straw rather than to put up with the production, storage and handling difficulties of feeding wafers. Secondly, in most cases the most profitable feeding system for the livestock farmer will be based on high-quality hay or silage, with dried grass fed together with rolled barley as the main supplement. In that case the hay or silage will provide the long fibre needed by the dairy cow, and there is then real advantage in *small* particle size in the dried grass, to allow the cows to eat the maximum amount of the hay or silage (p. 142).

There could, however, be a future role for the 'wafer' in the industralised feeding system advocated in Denmark by Wind. This complete feed would contain milled dried grass, cereals, minerals and chopped straw, the 'wafer' form being necessary to retain the long fibre form of the straw, needed by dairy cows. But considerable management advantages must be shown before this type of feeding will compete with that based on farm-conserved forages.

The mineral content of conserved forages

As with the major nutrients, energy and protein, the requirements by different classes of livestock for other nutrients, including minerals, are given in feeding tables. Other tables give the *average* contents of different minerals in different forage species; but it is difficult to predict the mineral content of a particular forage because this tends to be much more variable than, for instance, digestibility.

Trends can be noted—the generally higher mineral content of legumes than of grasses; the high calcium to phosphorus ratio in legumes; the low contents of sodium and magnesium in timothy— and these can indicate where supplementary minerals are needed. But mineral content also depends much on the soil on which the particular crop is grown, and it is a great advantage for the individual farmer to have a range of his crops analysed, to determine which minerals are likely to be deficient in the forages he harvests.

In most cases phosphorus, magnesium and sodium are likely to be deficient, but the amounts of these and other minerals which must be fed are best decided in consultation with a nutrition adviser, who will also consider the mineral content of the other feed supplements used. Again, where these include dried grass, a chemical analysis may by needed, although commercial supplements containing dried grass are likely to contain the appropriate minerals. Certainly, if he takes good advice, the livestock farmer should have little to fear from mineral problems as he increases the proportion of forage in his rations, and so becomes more dependent on feeds produced on his own farm.

*   *   *   *

This discussion of some aspects of the nutritive value of conserved forages has inevitably included information which will be repeated in the later chapter on feeding systems. But it seemed preferable for that chapter to be mainly factual, the systems being based on information outlined here, but this information not being essential for the description of the systems. It is certain, however, that this nutritional background will become increasingly important as livestock farmers change their methods of feeding to make more effective use of forages for profitable animal production.

Chapter 4

CROPS FOR CONSERVATION

MUCH OF the hay and silage made in the UK is cut from fields which are also grazed. Thus the particular crop species grown must also be suitable for grazing—particularly where this is the main method of use, and cutting is introduced largely to remove excess spring grass and to keep the pastures in good condition for grazing. In other cases grazing may be secondary to the main purpose of cutting for conservation; or all the forage may be cut, as with forage maize.

Clearly, where the principal use is grazing, the quality of the grass cut for conservation may sometimes have to take second place to the

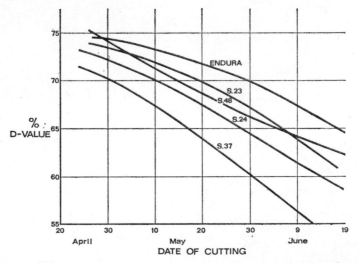

Fig. 3. The digestibility of several grass varieties during their first growth in the spring.

needs of the grazing system. But modern grazing management generally adopts either controlled paddock grazing or a two-sward system, in both of which paddocks or fields are cut at intervals, to give high yields of high quality forage for conservation. Except where cutting has to be introduced to remove an unexpected flush of grass, it should then be possible to combine cutting pretty effectively and predictably with grazing.

THREE IMPORTANT POINTS

Here, the recent information on yields and nutritive values, outlined in Chapter 3, will be of real practical use. For the pattern of

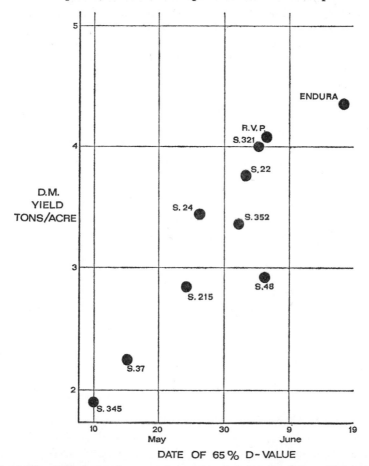

Fig. 4. The yield of several grass varieties when their D-value falls to 65% during first growth in the Spring.

growth and quality of S24, shown in fig. 2, is typical of the patterns for the many other forage species and varieties now in commercial use (fig 3 and GRI Technical Bulletin No. 8*):

(a) At a given date different forage species can differ markedly in digestibility: some species, in particular cocksfoot, tend to be less digestible than others, such as ryegrass and meadow fescue.

(b) Varieties within a species can also differ in digestibility; late-maturing varieties, such as S23 ryegrass, are more digestible than early varieties, such as S24, when cut on the same date.

(c) White clover is generally more digestible than the grasses, so that grass/clover swards are more digestible than pure grass swards; in contrast red clover and lucerne are rather less digestible than the grasses with which they are grown.

In making decisions on which crops to grow, yield as well as quality (digestibility and protein content) must of course be considered. An effective way of comparing crops is shown in fig. 4; this gives the yields of a range of grass varieties on the dates at which their digestibility falls to the same level (65% D-value) during first growth in the spring. The high yields of the ryegrasses, followed by timothy and meadow fescue, are clearly seen. As well as being higher yielding, the ryegrasses do not reach 65% D-value until the end of May or early June (when field conditions for cutting and wilting are likely to be better than in mid-May), by which time cocksfoot varieties have fallen to a low level of digestibility.

Fig. 4 also shows again the important point noted in (b), that different varieties within a species reach 65% D-value on different dates. Thus S24 ryegrass is at this level around May 26th, followed by S22 Italian, S321, RvP Italian and finally Endura, a recently-introduced Dutch variety, which does not fall to 65% D-value until June 16th, nearly three weeks after S24. This difference may seem small against the whole span of spring and summer, but it will be of increasing importance as more attention is given to the organisation of conservation systems. For if several fields on the farm are each sown to a different variety of ryegrass, a succession of first growths of forage of similar digestibility can be ensured over a period of several weeks—for example, S24 to be cut from May 22nd-28th, S321 from June 2nd-7th, and Endura from June 13th-18th, all of them at yields in excess of 3 tons of dry matter per acre.

ADVANTAGES OF PLANNED CROP SPREAD

This planned spread of crops could have great advantages. Firstly, if the whole acreage to be cut is sown to *one* grass variety, all the

---

* The forage yields given in this Bulletin are experimental, and 'farm' yields, quoted here, are likely to be somewhat lower.

forage will reach the same level of digestibility on the same date (except from paddocks grazed in April, for grazing slightly delays the fall in digestibility).

To harvest all the crop at a similar digestibility must put heavy demands on both labour and machinery. However, where only limited labour and machines are available, harvesting must continue over several weeks, so that the digestibility of the crop on the last field (and of the hay or silage made from it) will be much lower than on the first fields. In contrast fields sown to different varieties can be cut over a range of dates, giving hay or silage of fairly uniform digestibility and making full use of men and machines.

Furthermore, fields sown to a single variety are all at risk to the weather at the same time. This is perhaps not too serious with bunker silage, which is not very sensitive to the weather. But with hay or tower silage wet weather at the date at which the crop reaches the optimum stage of yield and digestibility means that all the crop has to be left to fall in digestibility until the weather is fit for cutting. With a succession of crops the weather risk is at least reduced— though one can still expect the occasional period of three wet weeks even in the British summer!

Finally, when conservation is combined with grazing, cutting a succession of swards produces a series of re-growths which themselves become available in succession for grazing—an advantage compared with the larger area of a uniform re-growth from a single grass variety cut at the 'optimum' stage.

Clearly this use of different varieties must be kept as simple as possible. In most cases different areas sown to two or three varieties should be enough to bring about a real improvement in management for both conservation and grazing. (This idea is of course not new: many farmers have sown fields to Italian ryegrass, which is ready for grazing earlier than perennial ryegrass or old pasture, and which gives some spread of dates at cutting time.) The bigger the conservation enterprise the more advantage there will be in ensuring a succession of forages for cutting. The most notable case is in grass-drying, where a steady supply of crops over a long drying season is a key factor in the economics of production.

Unfortunately drier operators are at present almost unanimous that ryegrass is a most unsuitable grass for drying, and so have not exploited the range of maturity types available in this species. But different fields separately sown with tall fescue, early and late cocksfoot, lucerne and spring-sown seeds are now used to provide a flow of crops for drying.

DIGESTIBILITY LEVELS

A comment on digestibility levels is important. In this discussion (and in fig. 4), a level of 65% D-value has been used to examine aspects of differences in yield and digestibility between different forage species; *this is not meant to imply that 65% D-value is the optimum level to be aimed at.* For, as noted earlier, the main use of these relationships is to assist the individual farmer to decide what to grow and when to harvest *for his own particular livestock enterprise.* Clearly there are several factors to be considered:

(a) The class of stock to be fed—dairy and beef cattle need more digestible feeds than suckler cows and store cattle.

(b) The type of enterprise—the winter milking herd can justify higher-digestibility conserved feeds than the spring-calving herd, which may be at a low level of production for much of the winter, but which may respond to a supply of top-quality material during steaming up, and in the weeks before grazing starts.

(c) The stocking rate on the farm—at high summer stocking rates relatively few acres will be available for cutting, and it may be advisable to take heavy cuts of relatively low-digestibility forage, and to buy in winter supplements. To date this has applied particularly to the smaller livestock farm, where the maximum number of stock has been kept to increase 'the size of the business', and on which winter self-sufficiency has taken second place to high stocking rate. Whether this situation will continue in future will depend on the price of bought-in feeds relative to the receipts from milk and meat, and on whether the use of dried grass can reduce the cost of supplementary feeding.

But within the overall picture the method of conservation to be used will have a major influence.

CROPS FOR HAY

Even with barn hay-drying facilities available, weather conditions in the UK are seldom suitable for cutting before the third week of May, and possibly rather later in the North. This means that crops are usually below 65% D-value before they are cut; much of the barn hay is therefore in the range 60-65%, although this is a very acceptable level. Cutting for field hay will generally be a week or so later, in the range 55-60%, unless late ryegrass varieties such as S23 are sown in order to delay the fall-off in digestibility until June; Endura, which need not be cut for hay until mid-June, could be of real interest here.

Another way of delaying the fall-off in digestibility is to hard-graze the sward up to the time when the ears have begun to move up the young stems. Grazing removes many of these ears, leading to a fairly leafy re-growth of high digestibility for cutting in June; but it must be accepted that the *yield* of this crop will be much lower than if it had not been grazed (perhaps not too unwelcome for haymaking) even if it is well-fertilised immediately after grazing.

Until the 1950s over a million acres, sown each year to Italian ryegrass/red clover mixtures, were cut for hay in June and July. These swards were often left down for two years, and ploughed for winter cereals in the second autumn. The acreage decreased rapidly from the mid-1950s under the combined effects of cheap fertiliser nitrogen (reducing the need for the clover), more intensive corn growing (with fewer break crops) and the greater incidence of clover stem-rot and clover eel-worm. The situation with all these could change in the future—with more expensive N, with the reintroduction of break crops into arable areas, and with the breeding of the tetraploid clover varieties, which are more disease-resistant than the older varieties. Such developments could lead to more acres being sown to Italian/red clover, much of this to be cut for hay.

But the greater part of our hay tonnage is still likely to come from old swards of mixed botanical composition. The aim with these, as with other swards harvested for hay, must be to cut considerably earlier than has been traditional, and then to adopt the methods described in Chapters 5 and 6, so as to make a medium-digestibility winter feed with the least risk of wastage or spoilage.

CROPS FOR SILAGE

A high proportion of the silage now made in the UK is cut as part of the paddock rotation on intensively-managed dairy or beef farms. Most of the forage cut is thus ryegrass, well-fertilised with N and containing little clover. As long as it is not cut in a very immature stage, ryegrass has a high sugar content and is well suited for silage-making; its use, both for grazing and for silage, is therefore likely to increase. Here, then is a particular advantage in using two or more different ryegrass varieties so as to get a spread of maturity for cutting. With smaller farms these different maturity types could well be sown on fields on several different farms, to allow co-operative use of the larger and more efficient harvesting machinery which the individual farm could not possibly justify.

Alternatively, the fields cut for silage can be completely separated from those used for grazing—the two-sward system. In that case a wider range of crops, which need not be adapted to grazing, can be considered; there is in fact increasing interest in these, as they can be introduced as break-crops in an arable rotation. This system is best adapted to larger farms, with grazing concentrated on the fields closest to the buildings, and conserved feeds cut from the arable fields further out.

The role of swards based on the tetraploid red clovers has already been noted for hay. These mixtures are likely to be equally important for silage, for with modern methods they make top-class material in both bunkers and towers. Particularly where the persistence of tetraploid clover is being exploited by leaving it down for two years, there may be advantage in sowing meadow fescue (S53) or timothy (S352) with the clover instead of Italian ryegrass; these grasses are rather less competitive than Italian, and they also maintain their digestibility a few days later, allowing cutting to be delayed into early June, when the red clover is that much more vigorous.

CEREAL SILAGE

After the introduction of tower silos in the 1960's, difficulties were experienced in wilting some crops cut before the end of May. There was thus interest in crops which mature rather later when wilting conditions are, on average, more certain. Timothy/red clover and grass/lucerne mixtures were used but there was also much interest in the use of whole-crop cereals.

This practice was not new—arable silage was sometimes made

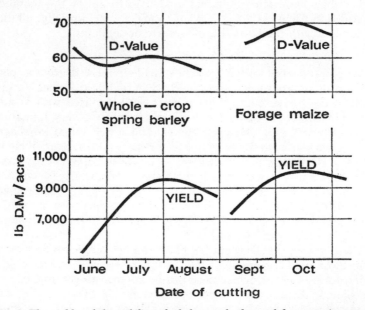

Fig. 5. The yield and digestibility of whole-crop barley and forage maize, cut on different dates (100 units N applied).   Data: A. J. Heard, GRI

before the war—but the airtight tower and improved machinery meant that crops of barley, wheat and oats could be efficiently harvested and stored at a more mature (higher dry-matter) stage then previously. However, these crops have not been as widely-adopted as was expected, and their use seems unlikely to increase. There were several reasons—yields in practice have not always been as high as experiments had indicated; the digestibility (D-value) of barley and wheat silage is seldom above 60% (fig. 5) and oat-silage is even lower; protein contents, at 7-9%, are lower than in most grass silages; and the subsidy system of the '60s meant that many crops did not attract the cereal subsidy when they were harvested as silage. With the higher cereal prices in the EEC it seems unlikely that many farmers will plan to make cereal silage, except perhaps from crops heavily-infested with wild oats. Further, in many areas the search is now for crops which will give a break in the extended growing of cereals; hence the interest in red clover, lucerne and maize.

FORAGE MAIZE

Forage maize is not new; large crops of the variety White Horse Tooth used to be grown, but harvesting was not mechanised, and quantities of effluent flowed from silos filled with this very late-maturing variety. The position, at least in southern England, has been transformed in the last 10-15 years:

(1) New varieties of forage maize, both from the Continent and locally bred, mature much earlier. By the end of September over half the dry-weight of the crop is present in the cob, and crop moisture contents of 25% or above make them ideally suited for silage. An NIAB list gives the latest recommended varieties.

(2) Improved growing methods, in particular the use of precision drilling, and of Atrazine (2 lb/acre) to control weeds, ensure higher and more uniform yields.

(3) Bird damage in spring can be a serious hazard, but will become less serious as more maize is grown; this is certainly the experience in North-west France, where the increase in maize growing has been spectacular.

(4) In many cases a few 'look-see' acres of maize have been grown; assuming it has survived the birds, the maize has proved most difficult to harvest with conventional harvesters—yet the small acreage has not justified special harvesting equipment. But as more acres are grown harvesters properly designed for the job will be purchased; again there is a good case for co-operative use of equipment, particularly as the exact date for harvesting maize is not as critical as in the case of the grasses.

This is because the yield/digestibility pattern of forage maize (fig. 5) is quite different from the grasses. In general both the yield and D-value of a maize variety rise to a maximum value, and remain fairly constant for a couple of weeks, in contrast to the S24 ryegrass shown in fig. 2.

By planting some fields to a medium-early maize variety and some to a medium-late variety, a further spread of harvesting dates can be achieved. Over this period there will be a considerable increase in the dry-matter content of the maize. This will not matter with well-sealed bunker silos, in which forage maize can be sucessfully stored over the whole range 20-35 % dry matter; but it is more critical with maize for tower silos, which must be left until it has reached 28-30 % DM before it can safely be loaded into the tower (although maize at lower DM can be used to 'top-up' a tower partly filled with grass silage made earlier in the year).

This must place some limitation on the suitability of maize for tower silage; certainly there cannot be a direct application to UK conditions of experience in N. America, where maize crops readily reach the 30-35 % DM suited to tower silos. On many soils delay in harvesting to achieve higher DM levels also brings the risk of soil damage from harvesting equipment as wetter autumn weather sets in. Unless there are major advantages to be gained from a fully-mechanised system of feeding from towers, storage in sealed bunkers seems the method which will be most widely adopted for maize silage.

ENSILED KALE AND CROP BY-PRODUCTS

Kale, pea-haulms, sugar-beet tops and vegetable waste are also used for silage. There seems little point in growing kale to make into silage when it can be grazed or cut and fed direct—particularly as kale often makes a pretty horrible silage. The other crops are by-products, and are only ensiled when animals are conveniently available to eat the silage in the winter. In practice they often make poor silage, with fermentation spoiled by high moisture content and soil contamination, but quality can be much improved by using an additive such as formic acid.

CROPS FOR DEHYDRATION

Grass drying in the early 1950s largely failed because crops were available for drying over only a short season—many driers operated less than 80 days in the year. The operators who remained in business—several of them now leaders in the industry—were those who learned to manage crops so as to keep on drying for 200 days or more.

Programmes to ensure a succession of crops for drying have been greatly helped by the information on crop digestibility now available. For profitable drying requires both a long drying season (to reduce overhead and labour charges) as well as a high-quality product to sell at an economic price. The quality objectives of high D-value and high protein content cannot always be combined in all the crops dried, but every crop should at least be either of high digestibility (e.g. Italian ryegrass in mid-May) or of high protein content (e.g. lucerne).

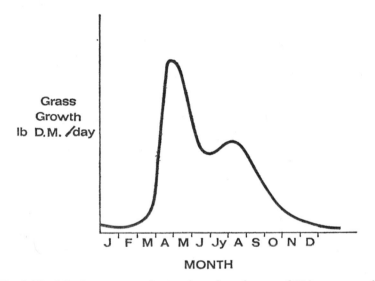

Fig. 6. The daily dry matter production throughout the year of S24 ryegrass, well-fertilised with N, and irrigated.

Most perennial forage crops have a basic growth pattern giving maximum yields in May and early June, less growth in summer and then a lift in yield in autumn (fig. 6). Some spread in growth can be gained by using different species and varieties, but the basic pattern remains. Thus the drier operator looks for other crops which complement this pattern. Examples are forage rye for early cutting (not an attractive crop because of its high moisture content); spring-sown grasses for cutting from mid-June onwards; whole-crop cereals or field beans for cutting in July (these crops are of rather low D-value, but are cheap to dry as their moisture contents are generally below 75% when they are cut); and forage maize for cutting in September-October.

Maize is of particular interest; although it has had little success in the UK, it is dried by many operators in France. This is partly because of the agronomic problems already noted with maize silage, but also because the French operators appear to have adopted the very fine chopping needed to dry this crop successfully. Forage maize is now well suited to the south of this country; it can give high yields of high D-value material for drying over a period of two months; and its low protein level can be made up by blending or compounding it, before sale, with high-protein grass or lucerne crops dried earlier in the season. From the evidence of its success in France it must soon be included in cropping programmes in many parts of the UK.

Clearly, to secure the steady flow of crops needed for drying demands a high level of field management, and the expertise needed is that of the arable farmer. The advantage of large-scale operation is also seen; for much the same skill is needed to plan a cropping programme for 200 acres as for 2,000 acres— and so the management cost per acre for the latter is only one-tenth of that of the 200 acres. Viable drying units of medium size will certainly be set up in basically grassland areas; but the advantages of planned crop sequences, of scale, and of the land being adapted to intensive harvesting, all indicate that the main tonnage production of dried crops will be from mainly arable farms.

Thus as the importance of conservation in the overall farming enterprise increases, so will the attention given to the choice of crops to be grown. At the one extreme the livestock farmer on permanent grassland will be concerned mainly with conserving the surplus grass which his animals do not graze in May and June; at the other, the grass drier operator may manage a range of crops to supply the drier over a long season. But both will need to give more attention to the stage of maturity at which they harvest their crops in order to produce feeds that will contribute more to winter livestock feeding.

Chapter 5

MOWING AND FIELD TREATMENT

IT HAS already been emphasised that whether grass is being made into hay, or wilted for silage or high-temperature drying, the aim must be to remove water from the herbage as quickly as possible, between mowing and removal for processing or storage, using a method which keeps loss of dry matter down to an acceptable level. This can be done only if the type of mowing and conditioning treatment used is related to the type of crop, the expected weather conditions, and the stage at which the crop will be removed from the field. The third point is especially true with haymaking, where the loss in the swath of most of the water present in the plant is accompanied by increasing brittleness of the leaf. The problem is accentuated by the high yields following heavy applications of nitrogen, which give both extra dry matter and water to deal with.

DRYING IN THE SWATH

Any method of complete drying in the swath will cause some loss of dry matter, and even during fine weather this will probably be at least 10%. As the minimum exposure time in the swath for a heavy crop is unlikely to be less than 48 hours, except for ensilage and some types of artificial drying, most cut herbage will be affected at some time by rainfall or heavy dew. During rain soluble nutrients are leached, leading to a steady decline in the digestibility of the remaining herbage. The loss of soluble nutrients increases with the number of times the crop is re-wetted, rather than with the total amount of rain, and the effect of a modest re-wetting from a low moisture content can be more severe than the effect of heavy rain on freshly-mown grass.

Given good weather conditions complete drying in the swath is a cheap and efficient method of haymaking, provided some form of mechanical treatment is applied at the time of mowing—or if this is impossible, within a very short time afterwards. Such primary treatment must be followed by further treatments at varying intervals. These cannot be dogmatically laid down, but as a general guide they should be applied whenever the top of the swath becomes appreciably drier than the centre or the bottom—always however with an eye to the weather.

REQUIREMENTS OF MOWING AND CONDITIONING EQUIPMENT

The key to success is the mower, or the mower-conditioner, often the cheapest item of machinery used in the conservation system. Whether it is the latest model available, or a secondhand machine bought and overhauled in the farm workshop, the requirements are similar if the mower is not to prove the weak link in the chain. Basic types of mowing machine change only slowly over the years. Although there are frequent minor changes in styling and mechanical features, the fundamental specification for the ideal mower or mower combination remains the same.

Briefly, the mower must cut the stubble cleanly to the desired height which, depending on crop and ground conditions, is likely to be between $1\frac{1}{2}$ in and 4 in. The swath which is formed should be evenly distributed along the field; for hay-making in particular it should have a large surface area in relation to its depth and should be open and well set up to allow wind penetration. Swaths mown for wilted silage should be in a form easily picked up and handled by the forage harvester, to ensure efficient chopping.

During this operation care must be taken that the mower is not damaged on soils which contain flints or stones. It should be able to operate continuously without blockage at a high forward speed in a wide range of crops, including those which are laid and which have a heavy and dense growth at the base. The design should be such that hard objects cannot be thrown upwards at the tractor driver, although it is sometimes difficult to prevent the ejection of missiles.

Its working rate must be adequate to deal with the total acreage of grass required daily for the particular system of conservation—probably in the range of 2-5 acres/hour, with an overall rate of from 65% to 75% of that figure for reasonably efficient operation. High output can be obtained by travelling at a high forward speed with a narrow width of cut, e.g. 5 ft to 6 ft, but performance with some types of mower is likely to be restricted by the discomfort the driver suffers if the tractor travels at much above 6 mph. Alternatively, a wider swath of 7 ft to 9 ft may be cut at, say, 2 to 4 mph; a possible disadvantage is that if the swath is then gathered to about 5 ft for subsequent handling the rate of drying may be substantially reduced. However, maximum outputs with pick-up forage harvesters are obtained in heavy dense swaths, and so if only a small amount of wilting is required, as for silage or dried grass, the cutting of a wide swath to achieve high outputs during harvesting is to be preferred.

The power requirement of the chosen mower must be well within the working capacity of the tractor. This may seem self-evident. But rate of work, standard of mowing, and recovery of crop dry-matter are frequently disappointing where tractors of less than 45 hp are expected to do a job which really required a medium or large tractor in the 45-80 hp range. Particular attention must be paid to this point in hilly areas, since claimed mower output often appears to be based on use in large fields on flat or gently undulating land.

MOWER-CONDITIONERS

Most mowers in current use cut the crop, but do not impart any physical treatment which will speed up the rate of drying. It is likely, however, that many more mowers or mower combinations will appear on the market during the next 5-10 years which will simultaneously cut and condition the crop.

There are several points to be considered in choosing this type of machine. Power requirement is likely to be fairly high; even if the unit is designed so that it does not chop the crop unduly when properly operated, lack of power for peak loadings will quickly lead to time-wasting blockages and will cause short chopping. This may lead to high fragmentation losses in haymaking; it may be less serious with silage, which needs less wilting, and where the flail harvester used to pick up the cut crop may have a slight 'vacuum-cleaner' effect. Even where a pick-up reel is used for silage, as on a precision-chop harvester, losses of leaf dry-matter will be less than with crops to be baled for hay. A really desirable feature of any combined mower-conditioner is that it should apply a more severe treatment to the base of the stems of the herbage than to the top leafy parts.

Maintenance costs of mowing equipment tend to increase with their complexity, and this is likely to be unavoidable, in addition to a higher first cost, with some of the more flexible high-output mowers. Regular maintenance and routine servicing requirements should be as simple as possible to avoid time wastage during the working day, although in fairness to the agricultural engineer it must be admitted that time lost through breakages is often more a reflection of the standard of out-of-season servicing than of machine design.

TEDDING THE CROP

As explained earlier, mowing the crop removes the growing plant's source of water and allows wilting to begin. However, the amount of water to be lost may be three or four times the final dried weight, and sap moisture which diffuses out of individual plants cannot be lost rapidly unless the swath is moved so that at least a drying wind, and preferably the sun's heat, can penetrate the swath and allow it to dry evenly from top to bottom. For this the crop must be moved at intervals. This is most commonly done with a tedder, the only piece of haymaking equipment, apart from the mower, which can be economically justified on many farms. The use of this machine, either at the time of mowing or as soon afterwards as possible, gives an early boost to drying rate. The same drying rate can never be achieved if the swath is left compacted and flat, as it is with the reciprocating cutter bar or by some disc or drum mowers.

Whether or not other methods of primary conditioning are used, tedding is still likely to be needed for secondary treatment. The ideal tedder is a multi-purpose machine, able to spread the swath thinly over the ground and to collect it together again into narrow rows. In that way half of any rain that falls is lost between swaths so that when drying conditions improve the uncovered stubble and soil dries out quickly, enabling the wet hay in the rows to be moved sideways on to dry ground.

CRUSHING AND CRIMPING

Tedders ensure that water is freely lost from the swath as a whole, and they certainly do much to reduce mould in hay. They do nothing, however, to condition the individual plant parts and so speed up the rate of moisture removal from within and, even more important, to increase the rate of drying of the thicker stemmy parts. One method of doing this is to crush the fresh crop as it is mown—or at the latest within 20 minutes of cutting, since treatment of a cut crop after it has lost its turgidity is less effective. *Crushing* is carried out by passing the crop between solid plain or ridged rollers. This flattens but does not shorten the stems, and has only a minimal effect on leaves, especially in lucerne and clover. *Crimping* involves passing the crop through corrugated skeleton rollers, which have some crushing effect but which also kink or bend the stems at 2-4 in intervals and may bruise the leaf. There may also be some shortening of the stems, while more severe damage to the leaf can be caused by 'burring' of the crimper bars which can occur on stony fields.

Both crushers and crimpers are designed to 'set-up' the swath for easy ventilation. In terms of improved drying rates and good dry-

1. Applicator for formic acid. Acid from the two 5-gallon plastic containers is fed by gravity directly into the chopping mechanism on the forage harvester.

2. Twin-drum rotary mower, 4 ft 6 in cutting width has 2 double-edged cutting blades per drum. Drum speed is 2,100 rev/min and forward speed can be up to 8 mph with 20 hp available at the pto.

3. 4-disc rotary mower with discs driven by gearing from below operate at over 3,500 rpm to give a knife speed of 300 ft./sec. Forward speeds of up to 9 mph can be obtained with a 45 hp tractor, depending on type of crop. The safety cover, together with a rotary divider on the outside disc to separate the mower from the standing crop, have been removed.

4. Flail mowers have a high power requirement, but leave a clean stubble under difficult conditions, and a swath which dries quickly.

matter recovery, these machines have much to recommend them. But they were developed mainly for lucerne and clover crops and have never achieved as high a measure of success in grass, chiefly because mechanical deficiencies have led to low overall output.

The more severe treatment of laceration and bruising, incorporated within some mowing machines such as modified flail harvesters and flail mowers, has the effect of splitting and shredding both stems and leaves, and leads to some short chopping which considerably speeds up the rate of loss of water.

During such treatment soluble nutrients are extruded to the surface of individual plant parts. These may become sticky and adhere to each other, and this can lead to some compaction and reduction in width of the swath. A disadvantage of the overall effects of this treatment is that even light rain can wash away much of the soluble fraction. Rain may also increase compaction, particularly of the partly-dried swath. This reduces air movement, the herbage then begins to degrade, and this can lead to decomposition and rotting.

Types of mowing and conditioning equipment

Cutter-bar mowers

Some requirements of the ideal machine have already been considered. The aim here is to discuss how far equipment now available, or likely to become available in the near future, will meet the needs of different conservation systems.

By far the commonest mowing device is the reciprocating fingerbar mower. But this can require much time daily on knife maintenance, particularly when used on stony ground, although it is a task that does not require much skill.

Standard of mowing may be poor in wet and heavy laid crops, and this can lead to 'bunching' along the swath following a succession of blockages, resulting in a heavy loss from uncut stubble. In standing crops without too much heavy bottom growth these mowers leave an even stubble length and as they cut the crop only once, they cause little or no fragmentation. Loss of dry matter directly attributable to mowing is therefore quite low.

This type of mower is comparatively simple, lightweight, and cheap to produce, and has a low power requirement of about 0·7 hp/ft width of cut. It is likely to remain the basic cutting unit on many farms for some years ahead. However, its forward speed is only in the range of 2 to 5 mph, and for a 5 ft cutting width overall output can vary from as little as $\frac{1}{2}$ acre/hr in heavy going, up to about $1\frac{1}{2}$ acre/hr in a standing ley.

Minor improvements will continue to be made which will affect ease of adjustment and use, as well as maintenance requirements and

c

performance. These are likely to include improvements to the drive by using hydraulics instead of the pto, elimination of the pitman in favour of a rocker lever, improved finger-styling, and lengthening of the knife stroke. Mid-mounted mowers, which leave the pto free, or rear-mounted mowers with through-drive power take-off, which allow simultaneous mowing and tedding or conditioning, should also be considered, for the cutting action of the mower does nothing to speed-up drying rate.

An alternative is the more expensive fingerless double-knife mower in which the multiple scissors-like action gives cleaner cutting under many conditions. It is less susceptible to blockage than the finger-bar mower, although 'bunching' along the swath can still occur in heavy crops. It is also less liable to damage by stones, because the high knife-speed throws them away from the cutting edges. A high forward speed, from 6-9 mph, is possible and output is up to $3\frac{1}{2}$ acres/hr, using 3-4 hp at the pto. A definite disadvantage is that maintenance standards must be high. Knife sharpening takes a long time, and requires special equipment with the skill to operate it if the potential performance is to be obtained continuously; against this from 15-25 acres can often be cut between sharpenings. To avoid delays some operators buy three sets of knives and have them sharpened by a dealer, but this adds appreciably to cost, while control of a critical operation is also taken out of the hands of the farmer.

Horizontal rotary mowers

The most significant recent change has been the development of horizontal rotary mowers with single or multiple rotors. The cutting action depends on the force of a fixed knife or swinging flail slicing through the plant stems, instead of the conventional method of cutting by shearing the plant between knives, or between knife and shear plate.

Single-rotor mowers of 5 or 6 ft cut, both in-line and offset, generally use two or three fixed cutter-blades, whilst twin-rotor machines use a variety of contra-rotating cutting mechanisms, including small swinging reversible flails and fixed cutter blades. These machines have fairly high power requirement but work reasonably free from blockages. Given adequate power they cut efficiently in laid and heavy crops. Knife maintenance is simple and up to 100 acres can be cut between sharpenings, depending on surface conditions. However, knives are easily damaged by large stones, and power requirement, which is as low as 10·5 hp at the pto for a 5 ft machine in good order, can more than double if blades become blunt or are badly set.

Rotary mowers give some conditioning by setting up the swath, but there is no worthwhile laceration or bruising to help improve drying rate; the overall effect is similar to that from one pass of a swath turner following a normal cutter-bar mower, provided that 'bunching' has not occurred during mowing.

The standard of mowing varies with design differences between makes. Because of the wide cutting width, the stubble length with single-rotor machines may be affected by ground conditions and can be very uneven on undulating fields. More flexibility is provided by twin-drum machines (photo 2) which may use an adjustable saucer-shaped disc at the base of a floating cutting head to control the height of mowing and to prevent 'scalping' of the ground. Picking-up crop over which the tractor wheel has run is difficult with some in-line machines, and with all types double-cutting of the previously-mown swath must be avoided, as this produces small pieces which can be lost during the later stages of haymaking. Overall performance can be improved greatly if these mowers are adjusted and used by a skilled operator.

Multiple drum and disc mowers

The greatest impact has been made by multiple drum and disc mowers (photo 3). These have the working advantages of single and twin-rotor machines, such as low knife maintenance and the ability to work in laid crops without blockages and 'bunching', plus an ability to follow ground contours without damaging the sward.

Fragmentation and stubble losses are generally quite low with the top-driven drum types, but a longer stubble is inevitable with disc machines as the drive is from a gear train sited below the cutter bar. Forward speed can be up to 10 mph, and continuous operation at 6-8 mph is practicable; the overall working rate is from 1-4 acres/hr, depending on the model, and it is often possible to mow from 2-2½ acres/hr over prolonged periods. These machines cost 2-3 times as much as normal cutter-bar mowers and require from 3-8 times as much power.

Power requirement of drum types is higher than of disc machines, which can be used at high forward speeds with a tractor of less than 45 hp. Both types leave neat and tidy swaths, but give little conditioning effect, while the thickest and wettest parts of the plants are set at the bottom of the swath with the more easily-dried part at the top. The alignment of plants tends to form a dense mat compacted in a narrow width and with a small surface area, and this restricts air movement. Tedding or similar treatment is therefore needed to obtain acceptable drying rates.

Tedders

Treatment of the swath as it is being mown, or very soon afterwards, and as often again as necessary until removal of the crop from the field, can be carried out by one of a wide range of different designs of tedder. These are constantly being up-dated and improved. Recent modifications add to the flexibility and efficiency with which they will do a whole range of jobs such as ridging into narrow swaths, spreading within swaths or over the whole field, windrowing for overnight protection or baling, and splitting down again if necessary.

The range of available machines is described in MAFF Mechanisation Leaflet No. 4. Except for the largest machines they are generally mounted on the tractor 3-point linkage. Single and double-row machines work at rates of 3-7 acres/hr, with 3 and 4-row machines giving outputs up to 10 acres/hr. A useful feature on some tedders is the facility to adjust the tines to suit the type of crop and the stage to which it has dried, so reducing loss of brittle leaf.

Combined mowing and conditioning equipment

Where very rapid wilting is required, there is likely to be much advantage in equipment which cuts and conditions the crop at the same time. This conditioning can be an integral part of the cutting action, or can be applied by separate mowing and conditioning mechanisms. Application of the flail principle to mowing for rapid swath wilting was first achieved on a large scale by the conversion of forage harvesters to *flail haymakers*. The most important alterations include fixing a suitable delivery chute, reducing the rotor speed, and removing the shear bars or other obstructions to minimise chopping of the crop. The use of 40 in and 48 in cut machines in this way is very successful as a first stage in the production of wilted silage, since the laceration and chopping leads to rapid drying in small compacted swaths, while the loss of cut crop is low because of the 'vacuum-cleaning' effect of the flail harvester used to pick-up the wilted crop. Higher losses can occur, however, when these machines are used for cutting crops for hay, because small fractions dry quickly, become brittle and are lost during baling. A few machines which lend themselves to such modification are still available. Provided they are *properly* adapted, and are used with a tractor of sufficient power to allow a forward speed of 3-4 mph whilst keeping rotor speed below 1,200 rpm, they provide a satisfactory mowing and conditioning treatment. They then produce an evenly-cut stubble with an acceptable level of chopping, which permits maximum advantage to be taken of fine spells on small acreages of hay.

This method of combined mowing and primary conditioning is more effectively carried out by the *flail mower*, specifically designed to cut and place the swath on the ground, and to provide laceration without excessive chopping (photo 4). These machines have large internal clearances to reduce obstruction to the flow of crop; some use fixed cutting knives, or flails which are restricted in their movement and have little fan effect, so reducing the no-load power requirement. Provided the cutting height is set correctly (including adjustment of any devices to reduce 'scalping' of the crop), flails are sharp, and both the rotor speed and forward speed are matched to power input to prevent the flails swinging back over the uncut crop, a tidy short stubble can be obtained without adverse effect on the rate of re-growth of the cut plants.

Flail mowers work particularly well in laid crops, but can leave a straggly stubble if used to cut with the 'lay' of the crop. Freedom from blockage and 'bunching' is a feature of the machines, but unless adequate tractor power is available there is a risk of blockage if bottom growth is heavy. Output, which is much influenced by crop type and power availability, varies from 1-3$\frac{1}{2}$ acres/hr for a 5 ft cut and up to 4 acres/hr for a 6 ft cut machine; power consumption at the pto is from 5-10 hp per foot of cut, so that 30 hp or more must be available at forward speeds from 3-6 mph. Since many farms have at least one high-powered tractor an output in most crop conditions of up to $\frac{1}{2}$ acre/hr for each foot width of cut should be possible.

Knife maintenance and repair requirements are much less than with reciprocating cutter-bar mowers, as damage by stones blunts rather than breaks flails. Loss of sharpness of course reduces cutting efficiency and increases power requirement, and there will often be a need for a major overhaul between seasons.

There is wide variation between different designs of machine in the amount of double-chopping and short-chopping which occurs. These variations influence the swath characteristics, some machines producing narrow tightly-packed swaths in which drying rate is restricted, while others leave a loosely-packed open well-ventilated swath. An undesirable feature of some designs is that a band of short-chopped material may be deposited on top of the swath; although this is easily picked up when collecting the grass for silage it can contribute to a high mechanical loss during haymaking. Flail mowing should in any case be restricted to grass, or to mixtures in which grass predominates, as the lacerating and chopping action causes severe fragmentation in legumes.

SECONDARY TREATMENT OF SWATHS

Swaths produced in this way will often need some secondary treatment before drying can be completed, especially after heavy rain has fallen on nearly dry hay, and consideration should be given to the width and type of the following machinery. Often a finger-wheel side-rake and turner will be the only requirement, but if a tedder is used it should be one which can be adjusted to provide a gentle handling action.

Hence flail-mowing has a place, certainly within the immediate future, for rapid mowing and wilting for silage and for haymaking in heavy-yielding grass crops. However, the high first cost of flail mowers, which are frequently heavy and cumbersome to use, their high power requirement, which often proves a limiting factor to output, and the high losses of crop dry-matter which *can* occur during haymaking mean that these machines are not ideal for mowing and conditioning. Flail mowers incidentally can be used for conditioning swaths cut with an ordinary mower to give fast drying rates; because chopping is negligible dry-matter loss is then no higher than for other methods of conditioning.

COMPOSITE UNITS

As an alternative to using the types of mowing equipment already discussed, the features of different types of mowing and conditioning equipment can be combined into a composite unit.

The most common equipment based on this principle combines a reciprocating cutter-bar mower with a pair of full-width adjustable conditioning rollers. Freshly-cut grass, rapidly removed from the area of the knife and fingers by an adjustable spring-tined reel, is fed evenly to the rollers. These may be of steel or rubber, usually with deep fluting on one or both rollers to improve crushing. The width of mower-crushers is commonly from 7 to 9 ft and they are designed so that the swath as it leaves the rollers can be gathered into a compact deep windrow, ideal for collection by a forage harvester after a short wilting period. For haymaking it is better to spread the grass into a full-width swath to give maximum drying rate; in fact if the swath is reduced to half the width of cut as it leaves the rollers the beneficial effect of crushing on drying rate can be lost.

This type of unit requires a tractor of 50 hp for operation at a forward speed of over 4 miles per hour. It should give an output of 2-4 acres per hour, but this can vary with the density of the crop being cut and with the incidence of blockage. Stubble height is usually higher than with normal cutter-bars. The cost of the mower-crusher is generally more than of a separate mower and roller conditioner, and the combined machine can only be justified on economic grounds if it is used on over 200 acres during the season.

The high first cost of such equipment (due in part to limited production), and difficulties of operation in some crops, have led to the development of alternative combinations. These are designed to speed up mowing without blockage and to produce an evenly conditioned swath, which will dry at least as fast as crushed and crimped herbage, and with no higher loss of dry matter.

One development has been the use of a series of free-swinging tedding flails, offset-mounted in a housing, to collect grass directly after mowing by a single-plate 4-blade horizontal rotary mower. This has produced a drying rate equivalent to immediate tedding but has not been widely used because single-rotor 5 ft mowers are not universally popular.

A second development, marketed in 1970, combines a belt-driven twin vertical-drum horizontal cutting unit with a forward-delivery spring-tined tedder (photo 5). The mowing drums are mounted with one slightly ahead of the other to give overlap in cutting and to deliver the grass obliquely to the tedding tines. Cutting blades, three per drum, are mounted on swinging flails so that they can swing backwards and upwards to clear most obstructions found in grass fields. Operated by a tractor of 35 hp upwards forward speed is likely to be up to 6 mph.

A COMBINED MOWER AND CONDITIONER

In view of the high standard of cutting which can be obtained with disc and drum mowers, the combination of these with an efficient roller crusher would have many advantages, and it is to be hoped that such a machine will be produced. The ideal combined mower and conditioning unit should of course mow at a high rate without breakdown and blockage, and should apply a treatment equivalent to crimping to the bottom thick parts of plants, at the same time giving a more gentle treatment to the upper leafy part. This allows more severe treatment to be given to the stems, which contain the most water and are the most difficult to dry, while the overall action of the conditioning unit should be to spread out the swath evenly and stand it up to dry.

A prototype machine, developed at NIAE (photo 6), and undergoing tests by a manufacturer, uses a forward-acting flail rotor for conditioning in combination with a reciprocating cutter-bar for low-power mowing. The cut crop is carried forward away from the knives, and upward and over the top of the rotor to the rear. Advantages of this arrangement are twofold. Rapid removal of the crop from the cutter-bar carries it away in a smooth flow and prevent blockages; because it passes through the rotor butts first the leafier parts are cushioned against the most severe crimping action, and this helps to equalise the rate of drying of the whole plant. As the flail is being used to condition only and not to mow, power requirement is comparatively low. The use of a reciprocating cutter-bar has the advantage of utilising the cheapest form of mower, but the whole unit can be equally effective with a twin-drum rotary mower. Likely power requirement at the pto for the 5 ft cut reciprocating mower and flail combination is up to 14 hp, compared with 14-28 hp required for a disc or drum mower, and over 30 hp when a flail rotor is used for cutting the crop.

Operating mowing and conditioning equipment

A number of general points apply to the use of all types of equipment. One of the most important steps to ensure trouble-free working with any mower is the rolling of grass fields early in the season. Heavy rolling will reduce breakages caused by stones and other obstructions, and is especially important with flail and rotary mowers.

It is often necessary to start mowing early in the day to cover the required acreage; but where a later start, say after 9 am, is possible on bright and windy days, the surface water which makes up much of the drying load early in the year will be lost much more quickly from the standing crop than from the cut swath. Particularly when flail mowers are used for haymaking it is an advantage to take headlands for silage, and then to work the field in lands about 50 yards wide; round-and-round mowing can cause a build-up of deep swaths at the corners, which are not easily handled by following machinery, even when operating diagonally across the field. These heavy wet patches dry very slowly. This either delays baling, or the wet grass is picked up too early, along with the drier parts of the swath, to form a nucleus for heating and moulding in the bales in store. The use of an inner swath-board on 5 ft and 6 ft reciprocating cutter-bar mowers is worth considering when the crop is to be wilted for silage, and picked up without further treatment with a flail harvester of 40 or 48 in width.

The importance of applying the first tedding or conditioning treatment either at the same time or immediately after mowing has been emphasised; indeed with machines such as crimpers and crushers any delay not only reduces effectiveness, but may also cause operating difficulties. Normal practice has been to apply the first conditioning treatment in the same direction as the mower, but particularly with some more recent machines it can be an advantage to work against the direction of cut of the mower, and this should be adopted where it is found to improve the standard of work.

Secondary treatment following a severe primary treatment such as flail mowing, which is accompanied by heavy laceration and chopping, must be strictly graded according to the type of crop and the stage to which it has dried, Ideally the swath should be left as mown until it is ready to be windrowed for baling, using only a finger-wheel rake; when additional treatment is necessary this should consist of turning and inversion with a finger-wheel rake so that the rolling action retains small pieces of hay within the swath.

The amount of crop cut for hay at one time should be matched to the available tedding, conditioning and baling capacity in such a way that each field, or part of a field, can be baled as soon as it reaches the required moisture content. This will often mean that after one pass the tedder should carry straight on with the next; in these circumstances it is possible with the highly-adaptable modern tedder to put up the hay into narrow compact rows overnight, so reducing the wetting effect of rain and dew, and to spread it out over the whole land area the following day as soon as the ground between the swaths has dried.

Large dense windrows should be avoided while the hay is still very green. If windrowing of two or more rows into one, ready for baling, is followed by rainfall the windrow should be split down into the original swath size as soon as possible, otherwise rate of drying will be very slow. Speed and efficiency of baling are closely related to the type of windrow presented to the pick-up reel. Maximum rate of baling is most easily achieved in a heavy and dense swath at a low forward speed, producing 36 in bales with five or six wads (strokes). But better-quality bales containing a larger number of wads, which dry out, handle and store better, can often be made by working at a higher forward speed in a lighter windrow.

LOSS OF DRY MATTER

Many trials and experiments have been carried out with the types of equipment discussed here. Although detailed results are not individually of value when making a choice of equipment, especially as they always relate to a unique set of crop and weather conditions, they can be very useful for comparative purposes.

Losses as uncut stubble are mainly of importance where a field is taken only for conservation, for when the re-growth is grazed much of the crop remaining after a field has been cleared will be eaten later by grazing animals. Rotary mowers operated at the correct speed can cut as cleanly as the reciprocating mower, but leave more stubble if the flails are blunt, or if they are operated at too high a speed in relation to tractor power. On the other hand, they can give lower stubble losses in laid crops, or where there is heavy bottom growth, provided they have adequate power to work at optimum forward and rotor speeds.

Loss of dry matter by fragmentation varies widely between

machines. In an experiment in which grass was collected at between 45% and 65% mc—a likely level for tower silage—the loss behind a flail mower was 6·0%, and with a modified flail harvester followed by a windrower as high a 28·0%, compared with a loss of only 2% with a finger-bar mower and conditioner. Failure to use the windrower to gather the crop together increased the loss with the flail harvester to 54% because a high proportion of the short pieces of chopped grass could not be recovered. Correct use of machines, particularly those using high horsepower and providing a severe lacerating treatment, can do much to reduce dry-matter losses from these high levels.

In another series of experiments, in which flail haymakers were operated at a wide range of forward and rotor speeds, mean dry-matter yield from the most severe treatment (low forward speed and high rotor speed) was only 80% of the yield for the best treatment, in which both rotor speed and forward speed were well-matched to tractor power output; up to one ton per acre was lost from a crop yielding 2·2 tons per acre where the secondary treatment was too severe.

Direct comparisons have shown that losses from flail and rotary mowing can be as much as 10% higher than from the control treatment of conventional mowing and tedding, although under some weather conditions losses behind the newer types of mower and conditioner may be significantly lower. Under most circumstances the yield obtained from mowing followed by crimping, crushing, or flail tedding will vary little from the yield with mowing and tedding.

However, misleading results can be obtained by assessing total hay yield at the time of baling because conditioned grass, whether crushed or lacerated, may have dried to a lower moisture content at baling than untreated grass, and may therefore lose less dry-matter in store. Thus the sugar content (water-soluble carbohydrates) of hay from different treatments has been similar at baling, but during storage has been reduced by half in tedded hay, by 35% in crimped but by only 20% in flail-mown hay. This selective loss of carbohydrate can significantly reduce the digestibility (energy value) of the crop. It is not too sweeping a generalisation to say that an apparently damaging effect of some conditioning treatments for hay has not always proved to be detrimental and has even been beneficial when measured in terms of yield and quality after a six-month storage period. Incidentally, a good correlation has been shown between the specific power requirements of a primary conditioning treatment and the rate of loss of moisture, but fragmentation losses can also increase with increased power usage.

Although higher drying rates can be obtained by increasing the number of secondary treatments, these must be applied with loss of crop in mind, and the advantages gained from quicker drying must be set against additional losses of 15-30% of the dry matter, which may occur with very leafy crops if the number of treatments is carried to excess, especially after flail mowing.

In practice a more severe *initial* conditioning treatment will nearly always reduce the need for *secondary* treatment by at least one pass through the crop. The cost of each pass is likely to be between 50p and £1 per acre, equivalent to $2\frac{1}{2}\%$ to 5% of the yield in a 2 tons/acre crop with a value of £10/ton. This possible reduction in total cost should be taken into account when evaluating mowing and conditioning machinery on the basis of dry-matter loss.

DRYING RATES

It is difficult to generalise about drying rates but certain patterns are likely to be repeated under a wide range of crop and weather conditions.

The fastest rate of drying is obtained by mowing with a 40 in-wide flail haymaker, although drying rate after some flail mowers is nearly as high. These treatments are likely to show a gain of from 2 to 4 days in field drying time compared with mowing and tedding, although this depends on the crop and on the type of tedder. Tedding while mowing gives an advantage of at least 48 hours compared with leaving grass untouched in the swath, while crimping, crushing or flail tedding gives a drying rate intermediate between tedding and flail mowing. In practice though it will not always be possible to bale flail-mown herbage as soon as these drying rates would suggest, because wet dense patches can remain after the main bulk has dried.

A slightly different situation exists where grass is mown and conditioned for wilted silage. Dry-matter loss is much less of a problem because the leaf never becomes really brittle (at moisture contents down to 60%) and the machine used for picking up the crop is less likely to cause loss of small fragments. Hence the aim is to achieve as much wilting as possible within a 24-hour period. Data were collected from a crop of Italian ryegrass cut in May during a fairly good drying period for the time of year, but interspersed with occasional showers. Mean time from mowing to picking up was 27 hours and the yield, about 4,500 lb DM/acre, did not differ significantly between treatments.

Even after this short time in the swath the advantages of tedding and conditioning compared with turning are clearly seen in terms of drying rate and dry matter at harvesting; the order of effectiveness of the other treatments is much the same as for complete drying in the swath (Table 6).

Regardless of the type and effectiveness of treatment, however, swath drying is still very dependent on weather conditions, and as the crop gets drier it becomes much more susceptible to damage from re-wetting, especially if it has been heavily conditioned. Losses can be much reduced by removing hay from the swath at a high moisture content and completing drying by fan ventilation, as discussed in the next chapter.

Wilting by heat treatment of standing crops

Present methods of accelerating loss of water from plants are either an adaptation or a development of cutting the crop with a sickle and tedding it by hand-fork. Thus there seems scope for the development of novel methods of treatment; one possible method for speeding up the rate of loss of water from the *standing* crop is the application of heat before mowing. Surface water is more rapidly lost from a crop through which wind can blow freely; this wilting of a crop, in which the cells have lost their capacity to retain water as a result of killing by heat treatment, will occur more easily before mowing than in a compact and relatively dense swath lying on damp soil.

Experimental work started in Holland in 1967, in which flame treatment was used to heat the uncut crop to 70°C; this damaged the epiderm and waxy covering of leaves and the top parts of plants, and caused some damage to the stems. However, the lower wettest part of the crop, particularly any dense bottom growth, received insignificant heating whilst the drier and more valuable leaf was scorched and easily ignited. Later development of a steam treatment, in which water is dispersed directly into the flame to produce a mixture of hot gases and steam, with a temperature below that which will ignite plant parts (i.e. 300°C), has advantages, apart from the reduced fire hazard. Because of more efficient and uniform transfer of heat to the crop, fuel consumption is reduced to half that required with a direct flame.

A field machine 17 ft long, 8 ft high and 8 ft wide, weighing about $2\frac{1}{2}$ tons and capable of dealing with a 20 ft width of crop at a rate of 4-5 acres/hour, (40 tons fresh crop) has now been constructed by a grass-drier manufacturer (photo 7). It can be used for other purposes besides crop wilting, including destruction of haulm and other unwanted crops, of particular value where residual chemicals present legal problems. Cost of fuel for treatment is about 19p per ton of dry matter, a negligible heat input in terms of the amount of water which can be removed from the standing crop. The additional cost of owning and operating such a machine has not been calculated. It will certainly be substantial, but this machine could still prove economical if it is used on a large annual acreage.

The possibilities of this approach in the production of hay and wilted silage are considerable. Provided that the crop can be left standing after treatment for a period from half a day to 3 or 4 days, depending on type and yield, water content at the time of mowing can be substantially reduced. Wilting for silage-making could be easily managed with this machine, and the cost of treatment might be little more than for mowing and subsequent conditioning. In practical trials crop dry-matter content was increased from 16% to 24% within 2 hours of steam treatment.

Perhaps the most promising application is for grass drying, because much of the drying load can be removed in the field more quickly and reliably than by normal wilting, and more cheaply than in the drier. In one experiment a crop of 84% mc was field dried to 71% in $2\frac{1}{4}$ hours, thereby reducing a drying load of 4·62 tons of water per ton of dried grass to 2·10 tons per ton. This would have halved the cost of fuel for drying, without the complications of dry-matter recovery involved in a mechanical juice extraction process, while the *capital* cost per ton of dried crop could also have been significantly reduced because of the increased rate of throughput.

More difficult technical and management problems are presented in using this kind of treatment for haymaking, because the crop cannot be dried completely as it stands. Reduction to a moisture content of between 50 and 60% can be obtained very quickly, representing a loss of from half to two-thirds of the water which must be removed before the hay is fit to put into store. Thus direct harvesting 12-24 hours after treatment might be practicable if a barn drier is used to complete the drying. But a heavy crop destined for field drying would need to be left for up to 4 days before mowing, and would then require tedding a few times to obtain rapid drying to less than 30% mc.

No information is available about the effect of rainfall on mown crops which have received heat treatment, but losses are likely to be similar to those when hay, which has been partly dried in other ways, is re-wetted.

Chapter 6

HAYMAKING

THE PROBLEMS involved in mowing and conditioning grass to make it dry evenly and quickly have been discussed. When the required moisture content has been reached, it is common, except in the heaviest crops, to put two or more swaths into one to obtain a high output from the baler or forage harvester. From this point onwards haymaking becomes a materials-handling problem. Except where some form of artificial drying is done, each further movement adds little or nothing to the value of the product but adds something to its cost.

THE HANDLING PROBLEM

The bale has the advantage of being an acceptable feed unit, in providing the daily ration for between 4 and 10 cows; it is also convenient to manhandle and feed, even at some distance from the store. Its density and symmetrical shape increase the capacity of transport, and bales take only half the storage space of loose hay. But the standard bale, 14 x 18 x 36 in, is really too small for individual mechanical handling, and too large for a free-flowing completely-automatic handling system. Movement from baler to store, and from store to stock, almost always involves manual handling, and no bale-handling system can be considered comparable with mechanised methods of silage handling.

To be economic any handling system must be well matched to the annual tonnage produced. Where the baler acts simply to pick up the hay, wrap it and throw it back to the ground to be moved into store by hand, machinery costs can be very low. In contrast, almost complete mechanisation *can* be achieved, but at a cost perhaps three to four times that of the baler; this cost will thus have to be spread over 500, rather than 100 tons of hay a year. Hay can of course be handled by other methods than baling; these generally involve further drying after removal from the field and will be discussed separately.

Baling hay

There are about 85,000 pick-up balers in use in England and Wales, and it would be quite unrealistic to expect a major change before the 1980s. So there is good reason to consider in some detail the most important points about balers and baling, and in particular to note some recent improvements.

Most balers in present use are of the slicing-ram type. In these the reciprocating piston, which presses the bale into the chamber, is fitted with a knife to cut through the flow of crop at the end of each stroke. Provided the knife is maintained in good order, and registers correctly with the shear plate on the chamber, neat bales, in which the 'slices' are cleanly separated, will be formed.

Bale density, although varying considerably with type of crop and moisture content for any one setting, can be altered over a very wide range, from 5-20 lb/cu ft, by adjusting the spring-loaded pressure plates. Wedges may also be used to increase resistance when the crop is very dry. Various forms of hydraulic control can reduce variation in bale density caused by fluctuations in moisture content— density is directly related to crop moisture content for a given set of conditions—but the advantages obtained have not proved economically worthwhile. Bale length, using a simple star-wheel and tripping-arm, can be varied over a range from 1 to $2\frac{1}{2}$ times the bale width on most modern medium and high density balers.

In contrast the swinging ram or press-type baler has a folding action, in which the successive charges are not positively cut. The bales produced are less tidy, and generally of lower density for a given crop condition, but have the advantage that air can penetrate the hay more easily and so allow better ventilation.

There has always been a small but keen following for rolled bales, which continues even though such balers are not easily available. Advantages of these bales are that they are almost weatherproof when left singly in the field, by contrast with the other two types which give protection only against dew and light rain. They also retain leaf much better. Handling and stacking can create problems, but these have usually been solved to the satisfaction of confirmed users.

IDEAL REQUIREMENTS

There are many minor variations between makes of baler, but the ideal requirements are common to all. The pick-up should be wide enough to deal with a heavy crop, the flared type being capable of taking in a swath of over 5 ft. To get a clean pick-up of all the crop, the combination of narrowly-spaced pick-up tines, a freely-floating counterbalanced pick-up and a spring-loaded crop guide is likely to be the most effective. The use of a flat plate to replace the normal tined crop-guide has helped with the very short 'fluffy' hay often produced by flail mowing.

The crop must pass in a clean unobstructed flow from pick-up to bale chamber. Efficient packers, which can be easily adjusted, are essential if good-shaped bales are to be formed. If the crop is fed either too far or not far enough into the bale chamber, the bales will curve away from the too-dense side; apart from the nuisance of strings becoming detached, irregularly-shaped bales reduce the output of many handling systems.

There has been a progressive increase in baler plunger speed, up to 80-90 strokes per minute at a standard pto speed, which leads to smoother running, especially when the plunger is mounted on sealed ball-bearing rollers. Clearly this increased speed is an important factor in obtaining high output, but the uninterrupted operation is also assisted by provision of a heavy flywheel, which smooths out surges and reduces the peak power requirement cause by variation in swath density.

The baler knotters should be able to work with both sisal and polypropylene twine, although minor adjustments, for example to twine tension, may be needed.

Another important consideration is the provision of a strong draw-bar—firmly attached to the chassis so that it cannot possibly cause distortion and is suitable for towing any bale-handling equipment which is chosen. Wherever possible, bale-grouping devices which throw an uneven load on the bale chamber must be avoided. A rapid and easy method of moving the baler from the 'transport' to the 'working' position, without the driver leaving his seat, can save both time and effort.

For high performance a baler requires good annual out-of-season maintenance as well as daily attention, particularly to lubrication of knotter and needle assemblies. There has been a gradual decrease in the total number of grease points, in favour of self-lubricating sealed bearings; some lubrication systems, when topped up, are effective for up to 20,000 bales.

The technique of operating a baler to produce well-shaped regular bales of reasonably even weight can only be learned in the field. Most of the *technical* data required will be included in a good instruction book, but there are a number of general points which are worth noting.

5. Combined off-set 5 ft rotary mowing and tedding unit. Wheels can be individually adjusted with screw-jacks to control cutting height, and hydraulic rams are used to lift the machine out of work.

6. 5 ft prototype reciprocating cutter-bar mower and flail conditioner. The forward acting Y-shaped flails remove crop from the area of the cutter bar, to prevent blockages, and also give a more severe conditioning to the lower stemmy parts of the plant than to the leaves, resulting in an increase of up to 40% in drying rate, compared with mowing and tedding, for only a small increase in dry matter loss.

7. Thermic crop conditioner uses super-heated steam to "kill" the standing crop and so reduce moisture content before it is mown for conservation. It carries 60 gal of oil and 600 gal water and can operate for 30 min before tanks are re-charged.

8. Random bale collector.
Sledges of this type collect
15-20 bales at a time, which are
dropped as windrows across the
field.

9. Wheel-mounted bale collector for random grouping of up to 20 bales at a time in
windrows, from which they can be rapidly hand stacked to form unit loads.

10. Automatic bale sledge. Models are available to form bales into flat layers of 8 or 10 which are automatically discharged into a uniform heap, ready for handling by a suitably designed loader.

DENSITY ADJUSTMENT

The baler must be adjusted frequently to both the type and the moisture content of the crop; otherwise bales will vary from very heavy in crops of moist leafy hay to very light in stemmy open crops of very dry hay—with bursting of the strings at the one extreme and collapse of bales from within the strings at the other. Adjustment will be needed at sunset, whether or not dew starts to fall, and there is a particular problem on the headlands of fields near hedgerows and woodland.

Hay is normally baled when its moisture content is between 20 and 35%. At the most likely level of 25-30% mc density should be about 10 lb/cu ft, giving a standard bale weight of about 50 lb. With drier hay density will be about 7 lb/cu ft, and bale weight 35 lb, and with wetter hay (over 30% mc), 13 lb/cu ft and up to 70 lb weight. An extended chute, connected to the rear of the bale chamber to deliver bales to a trailer, can increase bale density and weight.

When hay is baled at moisture contents above 35%, e.g. for barn-hay drying, density increases rapidly unless tension is slackened; a good guide is that it should be possible to thrust half a hand into the side of each bale after it has been released from the chamber. Wet bales can often only be kept at a reasonable weight, say under 70 lb, by reducing length to 27-30 in; this shortening has the further advantage that the bales are more likely to remain intact within their strings after drying.

Earlier mention has been made of the number of wads (strokes) per bale, which is directly related to the rate of feed of the crop and the ram speed. Bales with up to 20 wads, made from a light swath at high forward speed, dry out well if left stacked in the field, but can become rather loose. For barn-hay drying the ideal bale has 10 to 14 wads, whilst well-formed bales made from a large windrow at high output will contain 8-10 wads.

BALER OUTPUT

Although baler output varies widely according to crop and weather conditions, some guidance is required to the acreage of hay which can be made at one time, and to the capacity of the ancillary equipment needed to move bales from field to store. Very high rates of work, between 10 and 15 tons/hour, have been recorded over short periods with balers with high ram speeds and enlarged bale chambers, e.g. 18 x 16 in instead of the normal 18 x 14 in. But for most types of baler overall output varies from as little as $3\frac{1}{2}$ up to 10 tons/hour, and a general on-farm rate is $2\frac{1}{2}$-3 acres, or 6-7$\frac{1}{2}$ tons/hour. This rate, of 250-300 bales per hour, is well below the potential we can expect from the latest types of baler; but it is far in advance of many bale-handling systems. Without mechanisation a sizeable gang of men, working at about 45 bales per man-hour, would be required to clear bales being produced even at present rates of output.

BALE HANDLING

The mechanisation of bale handling from field to store can be divided into the 5 stages of grouping, loading on to transport, transporting, off-loading from transport, and loading to store. Developments in these have been relatively slow; minor improvements have taken place within the limits of a few basic ideas, and some attention has been paid to the relationships between one stage and another. In the 'ideal' handling method the bales would be taken directly from the baler to the store, with the output in each stage of the system matched to the succeeding stage, to ensure that one weak link did not adversely affect overall output. But various constraints and general considerations must be taken into account.

Often the most important is the need for a reasonably weather-proof temporary holding stage in the field, in which heaps of bales can lose water before final storage. Wherever possible, the size and shape of these heaps should form an integral part of the handling system, and the formation of groups of bales should not be directly tied to baling if this reduces baler output. The overall length and size of any handling equipment must be related both to field size and topography, and to building size and layout. Accessibility around buildings often limits the type of equipment which can be chosen. With some handling systems, based on large unit loads, effective storage space can be reduced if the dimensions of the load are badly matched to the size of the building, so that the space left is too small to take further bale units.

The first priority in choosing equipment is to decide how many bales (or tons) of hay will have to be handled in a specified time so that deterioration between baling and protection in temporary or permanent storage is avoided; for the feeding value of hay can be drastically reduced during this time. Timeliness is therefore a primary consideration, but clearly the rate of handling must also bear some relation to the available labour force and to the amount of capital which can be invested in trailers and specialised equipment.

Both low and high capital cost systems may have a high rate of output per man-hour, but when hay bales are at risk it is the total rate of working, rather than absolute labour economy, which may be of most importance. In short an efficient one-man-operated mechanised system may be too slow for the farmer with a large acreage, and too costly for the farmer with only a few bales to handle. So there must be a balance between equipment cost, labour requirement and output, and the trend is likely to be towards low-labour, high-capital systems which will clear a large tonnage with few men. This may well mean a swing to co-operative ownership or contract working, to justify the purchase of expensive equipment.

With the exception of round bales, the single bales which are left by the baler are easily damaged by heavy rainfall, and can also give a patchy re-growth over a wide area if they are left for more than a few days. A partial solution to both these problems is to use a fore-end loader to make heaps of bales in rows down the field or on the headlands. Covering the tops of the heaps with plastic sheeting, if possible raised slightly clear to avoid condensation, gives almost ideal field conditioning; this practice is much more common in the North of England and in Scotland than elsewhere in the U.K.

HANDLING SINGLE BALES

Single bales can be lifted directly from rows and loaded to trailers with an automatic pick-up elevator, at up to 180 bales/man-hour, while with a tractor-mounted hydraulically-operated swinging arm, bales can be tumble-loaded by one man at over 400/hour. With these methods a team of three men can haul and store close on 200 bales an hour, which at 70 bales per man-hour is double the working rate of most hand-loading systems.

The use of extended chutes and throwers, attached to the rear of the bale-chamber, which convey bales directly into the trailer, allows very rapid clearing of a field, provided enough trailers are available. Linking loading directly to the baler can, however, reduce baler output by as much as 25% because of a progressive fall in the sustained rate at which men can stack bales on the trailer. Even with random throwers and tumble loading much time can be lost with this lengthy equipment in making large loop turns in the field, and in hitching and unhitching trailers; as a result the *overall* baling rate can be reduced to 200 bales/hour.

Wherever possible, the handling and transport system should be planned as a completely separate operation from baling, so that the baler can work at its maximum output. In particular this means that any manual work is not geared directly to the rate of output of the baler. Manned sledges are therefore being displaced by windrowing, with random bale-collectors and automatic grouping devices.

Very rapid collection from such windrows, at rates of over 300 bales/man-hour, is possible by using a large low-platform trailer on to which single bales are loaded manually; the ideal working height is between 24 and 62 in from the ground.

Random-grouped bales can be built into field heaps by hand at rates up to 480 bales/man-hour. This labour can be a worthwhile investment in terms both of the weather protection provided by these heaps, as well as the unit loads they form for subsequent handling systems.

The bales from these heaps can of course be loaded singly, using an elevator, and two men can load a trailer at about 190 bales/man-hour. But it is much better, once bales have been formed into a group, to handle them as a complete unit from field to store, and even from store to stock, using tractor-power rather than man-power.

GROUP HANDLING SYSTEMS

Many different combinations of equipment may be included in a complete handling system and, provided the output of the successive stages is well matched, or at the worst a bottleneck in one stage does not reduce the rate of working in another, bales can be quickly cleared from field to barn.

With a manned sledge, on which heaps of from 8-20 bales are built as the first part of the system, the stacking rate can range from 200 bales/hour, with bales of over 60 lb weight, up to 350 bales/hour with 25 lb bales. Baler output can fall by up to 20% because sledges restrict forward speed to 3-4 mph, and cause slowing down when the heap is deposited; with heavy bales the working rate may also be progressively reduced by tiredness of the operator.

Thus unmanned sledges and collectors are preferred. There are two main types, random collectors, and fully automatic bale-sledges, which form the bales into groups that can be picked up by a tractor loader without manual handling.

Most random collectors (photo 8) are metal cradles which group up to 20 bales, depending on the height of the sides, and enable the tractor driver to form neat windrows. They are cheap and uncomplicated, but bales tend to be handled roughly, and particularly on flinty soils, strings can be broken or pulled off as they are dragged along. Various types of wheeled carrier collect the bales via a chute either on to a platform or into a V-shaped container (photo 9). Such equipment costs up to three times as much as the simple ground collector but it has advantages in reducing bale distortion and damage, in leaving a more compact heap for man-handling, and in giving less trouble on sloping land. It is also easier to see quickly when the knotter had failed to operate, so avoiding the production of large numbers of untied bales.

The aim of most inventors, and especially of inventive farmers with a real understanding of the problem, has been to eliminate completely the need for hand-formation of unit loads. Many attempts have been made to produce fully-automatic sledges, which build a heap of bales suitable both for weather protection and for handling. But as yet no machine has been made and marketed which can place bales directly from the baler into stable units of 16 to 20 in 4 or 5 layers, the unit most suitable for handling with buckrake and side-gripping types of loader.

AUTOMATIC SLEDGE

One type of automatic sledge (photo 10) forms flat layers of 8 or 10 bales without any hand labour. Unless the bales are to be carted directly to store, they must then be stacked into heaps of from 16 to 64, using an impaler loader, to provide weather protection; this can be done at up to 800 bales/hour. An alternative system uses a sledge which places two pairs of bales end-on to each other and then places another two pairs on top to form a heap of eight (photo 11); another sledge builds heaps of six, with two layers of three bales end-on to each other. Bales in these various forms can be left in the field to weather for a short time.

A wide range of loaders, varying from modified buckrakes with built-up carriers to units with mechanically- and hydraulically-operated side-gripping arms are available to handle heaps containing from 8-20 bales. With a few loaders the heap must be built in a particular form, or to a special shape such as a pyramid which offers better weather protection than the flat-topped heap.

The rate at which trailers can be loaded and bales hauled to the barn depends very much on trailer size and positioning and by the proximity of heaps to each other. Heaps of 16 bales can be loaded with buckrake-type loaders at up to 460 bales an hour if two men are stacking on the trailer, i.e. about 150 bales/man-hour. Squeeze-type loaders, which set down blocks of bales without manual stacking, can handle between 350 and 600 bales/hour, and impaler loaders (photo 12) also operated by one man, can load up to 540 bales/hour, or even more with trailers large enough to take four units of eight in each layer.

Direct haulage to the store on the front and rear-end of a tractor offers considerable economy of labour and a high rate of working over a round journey of up to a mile. However, using a tractor as a pack-horse is limited in scope by legal and safety regulations, especially on public highways.

Loading into store

Mechanisation of bale handling has concentrated on clearing fields as quickly as possible to reduce weather risk. At the store reduction of hand labour has been limited mainly to the use of an elevator to move the bales from ground or trailer level. Labour cost has been high, and work rates have seldom been matched to the field collection and transport handling system. In fact up to half the total labour cost of carting frequently goes into loading to store. Provision of extra trailers or the dumping of heaps of bales on the ground can prevent this final stage from seriously restricting the rate of removal of bales from the field.

Rate of hand-loading bales to an elevator depends on how easily they can be picked up; bales tipped in random heaps can be handled by one man at about 400/hour, and at up to 520/hour if they are taken either from trailers or from evenly-set-down rectangular heaps. With a further two men needed to load in the barn the respective outputs are 136 and 173 bales/man-hour.

Where layout is convenient front-end loaders are being more widely used for loading into store. Optimum utilisation depends on the barn size being well-matched to unit-load dimensions, and the loader should be able to stack bales at least 14 layers high. Using an impaler-loader to handle flat 8's, nearly 400 bales an hour can be unloaded from a trailer and stacked into a barn by one man.

OVERALL SYSTEM OUTPUT

While the rate of working of individual parts of a system can be spectacular, the overall output from field to store is often very similar with different systems—although output per man-hour can be considerably improved by use of the more expensive handling equipment. The figures given below indicate average rates of working, although much higher outputs have been obtained in isolated cases.

With a simple low-cost system, comprising a bale-collector and 4-wheel trailer and elevator, 100 to 120 bales/hour can be hand pitched and loaded by three men, transported about half a mile and unloaded by bale elevator to the stack; output is 33 to 40 bales/man-hour. By substituting two special-purpose low-loading trailers in place of the slightly cheaper 4-wheel trailer and field elevator, two men can handle up to 125 bales/hour, increasing output to 62 bales/man-hour. From 85 to 100 bales/hour can be moved by direct haulage over the same distance, using tractor-mounted front and rear bale-carriers. Two men may be needed for part of the time, but output is likely to be at least 50 to 60 bales/man-hour. Using more expensive specialist equipment, such as an automatic sledge and impaler-loader with large 4-wheel trailer, one man can load, transport and stack from 95 to 100 bales/hour.

NEW DEVELOPMENTS IN HANDLING

Most ideas on improving the rate of handling of baled hay are still based on the principle that the hay will be formed into stable units, which can be handled as a block throughout each stage into store, and in some cases for feeding out to stock. They depend on the grouping of bales, with cross-sectional areas of up to 16 in x 18 in, which were developed for manual handling as single units, although some work with random handling has been done with bales of 12 in x 12 in cross-section.

An American development is a pick-up device and a loading platform, either towed or self-propelled; bales which have been dropped singly, with the cut edge uppermost, are picked up and automatically formed into a block of either 55 or 120 (photo 13). This load can be transported at high speed and set down to form part of a stack by turning the platform through 90°; with it one man can handle up to 2,000 bales a day. The high cost of this equipment must restrict its use to large farms, or to co-operative ventures or contract working. It is also suited only to dry bales, and it may be a disadvantage under UK climatic conditions that it does not incorporate an intermediate holding stage for field conditioning.

Bales formed into unit loads of from 8 to 20 are often unstable, especially if handled more than once, and systems which tie them together positively can help speed up handling rates by reducing time lost when heaps collapse. The ideal would be an automatic sledge in which stacking and banding are done as a single operation; farmers and others have been working on such equipment for at least 10 years and no doubt a suitable unit will eventually be developed. However one existing system goes some way towards this aim, although manual formation of groups of 4 bales is required before the machine picks up and forms the unit of 20 bales (photo 14). Operation of the banding unit is at present also manual, so that to reach the claimed grouping rate of 400 bales/hour by one man the tractor driver would have to leave his seat 20 times an hour.

However, it is likely that a system can be developed in which banding will be carried out by a second man whose main task is grouping the bales into flat fours. After banding the bale blocks can be rapidly loaded into trailers and unloaded at the barn using a hydraulically-operated clamp-type loader (photo 15). Independent estimates of the daily output of this method under varying conditions are not yet available, but overall working rate, including handwork, is likely to be very competitive with other unit-load handling systems. It may also be an advantage that gaps between the blocks of bales allow free ventilation of stacks and, provided the bonded units hold together, feeding-out to stock should be very easy.

LARGE BALES

The tractor fore-loader has a lifting and carrying capacity which equips it to handle a single large stable unit, and this has directed attention to the production of large bales. Very large round bales weighing up to ½ ton have been made with balers developed in Australia, where, apart from handling to store, the facility has been used to retain hay in the field for feeding during drought; experimental equipment has also been developed in the mid-West of America, to produce round bales weighing up to 1 ton, and this may eventually be available for on-farm use.

In the UK in the early 1960s a Scottish farmer, Ian Sutherland, developed the idea of producing rectangular bales weighing about half a ton, following previous work in which he had palletised normal bales for handling. Experimental work was carried out at the NIAE in 1966 and 1967 to evaluate the handling, drying and keeping qualities of bales weighing from 1,100 to 1,990 lb, and measuring 6 ft x 5½ ft x 4 ft. Bales were produced in a static press at moisture contents ranging from 20% to 35%, and with densities from 8 to 15 lb/cu ft. Although these bales heated more than comparable stacks of 16 small bales, good hay was made after a period of field conditioning; it was also possible to dry them by artificial ventilation.

Bales of 130 cu ft and suitable for storage are likely to weigh between 1,100 and 1,300 lb. Studies have shown that they can be loaded to trailers at up to 60 bales, or 30 tons an hour and that, using large trailers, one man should be able to move them from field to store at a rate of at least 5 ton/hour. More rapid rates of work are possible using front and rear-mounted tractor carriers if farm buildings layout is suitable (photo 16).

Large bales are stable when loaded on trailers, as long as the bale height is less than either its width or length. They should not need roping, nor should banded packs of 20 bales.

As with fully-automatic sledges, equipment to produce large bales is not at present offered for sale. An application for patent was filed in 1967 by two Gloucestershire farmers—Pat Murray and David Craig—and the complete specification of an automatic-tying pick-up baler, which incorporates a novel method of packaging the hay in a series of small bundles within the large bale, was published in August 1971. Many thousands of bales have been made and handled with the prototype (photo 16a), but the future of this and other new handling ideas will depend very much on their performance in the hands of farmers and tractor drivers who have not been responsible for their development.

HAY CUBES AND WAFERS

An attractive alternative to bales of any size is the production of cubes or wafers weighing only a few ounces, and with a density three or four times as high as baled hay. The field machine at present available will only produce a cube, $1\frac{1}{4}$ in x $1\frac{1}{4}$ in x 2 in, from immature lucerne hay which has been dried in the swath to below 10% mc. This limits its use to arid zones, as in California. Experimental machines which will operate with grass crops over a wider range of moisture content to form rolled wafers of 2 in diameter, with a bulk density about 25 lb/cu ft have been produced in both America and Germany.

After 12 years of development many of the early mechanical problems, such as susceptibility to uneven feed rate and the production of very hard cores, have been largely overcome. Enough technical information is now probably available to make a production machine; whether this is made will depend mainly on commercial and economic considerations, for such a machine would compete directly with the pick-up baler. Hay packages made in this way would probably require either the use of an additive such as propionic acid, or forced-air conditioning, to prevent heating and moulding; thus, as with other forms of bulk handling of hay, the complete system of conservation might well need to include the provision of a storage drier.

Barn hay drying

The value of mechanical treatment of grass in the field, to improve evenness of drying between leaf and stem and to speed up the rate of loss of water, has been emphasised as a means of reducing loss of nutrients. However, drying to a low moisture content in the swath is difficult because, as the hay approaches equilibrium with atmospheric humidity, rate of loss of water is reduced; over-dried leaf is lost but stems remain succulent. Consequently the time of removal from the swath becomes a compromise between accepting a moisture content which is too high for safe keeping in store and leaving the crop in the field to suffer further leaching and physical losses.

The position is partly remedied by grouping bales and leaving them to weather in small heaps or stacks, but a more certain way to prevent losses from heating and moulding is to apply some form of fan ventilation, generally referred to as barn hay-drying.

Much of the development work into techniques of barn hay-drying in the past 20 years has been done by the Electricity Supply Industry which has made available detailed advice on installations and techniques*. Yet despite much advice from manufacturers of

* Green-Crop Drying. A guide to the practical design of installations. Farm Electrification Handbook No. 15, June, 1970.

fan and heater units, and of specialist drying equipment, as well as from Mechanisation Advisers of ADAS, the various systems have been slow to gain farmer acceptance, and it is probable that less than 4% of all the hay in the UK is made in this way.

The advantages to be gained are not in doubt. They include improved yield of dry-matter per acre—up to 15% higher than with normal field methods of production—and the greater feeding value of the recovered crop (Table 2—Appendices); hay can be improved from maintenance only to a feed which can support the production of a gallon or so of milk. So there must be some good reasons for this slow development.

Briefly the low output of some batch systems, coupled with the need for double man-handling, gave the impression that barn hay-drying was suited only to small farms producing perhaps 30 to 40 tons a year. Handling costs were high through lack of mechanisation, and where long unbaled hay was dried the need to spread it very evenly in the drier was a particularly time-consuming operation. Poor design of driers and inadequate provision of airflow also often led to uneven and inefficient drying. Because of the low throughput of some drying units, grass was frequently not mown until well past its optimum growth stage, so that despite the added expense of handling and drying the feeding potential was no better than swath-made hay.

Fortunately enough farmers made hay which gave sufficient extra production to convince them and others that this method of storage was worth developing and improving. The recent steady increase in the numbers of driers has of course been helped by improvements in techniques and equipment; but it has been fostered to a large extent by farmers who have sought to obtain better returns from their grass conservation programme.

SYSTEMS OF DRYING

Artificial drying can be divided into two broad categories. The first, *conditioning*, is simply a method of ventilation in which enough air is passed through the crop to prevent heating, and at the same time to give a gradual removal of a small amount of water—less than 6 cwt/ton of dry hay—over a period of 10 to 15 days. This will keep respiration losses to a minimum. It also helps to avoid conditions of natural convection, aided by heating, which removes water from the crop in one part of the store and deposits it in another part, usually in the top layers. Understandably, conditioning has been regarded as a process which can *be attached to the end of an existing method* of hay production, to reduce the risks of excessive loss *when the weather is very bad;* thus it is particularly suited to dealing with crops of 35% mc or less. Drying will continue so long as the relative humidity of the air is less than the equilibrium value equivalent to the hay moisture content (Table 7). Hence unheated air of less than 90% relative humidity will not only keep hay cool, but will also dry it to below 30% moisture content. The usual recommendation is to ventilate continuously for the first few days to keep the hay cool, and thereafter to switch off the fan at night and to turn it on again whenever air conditions are favourable.

But it is also possible to use the heat produced by respiration, whilst the fan is switched off, to help dry the crop; ventilation is then restricted to preventing the hay heating to a dangerous level. However, the savings in fan power must then be set against the in-store loss of crop dry matter, which could be well above the 4 to 5% normally expected when conditioning from 30% mc.

The second broad category is *drying*, in which sufficient air, as well as added heat, is used to remove between 6 and 14 cwt of water for every ton of dry hay produced at 15% mc. Moisture content at loading, depending on the chosen system, will be 35-50%, a level at which hay will not store satisfactorily, without ventilation, even if weathered in the field for a considerable time. Therefore drying becomes *an integral part of the haymaking system;* without it, a storage loss of at least 15% could be expected.

Drying must take place quickly enough to enable the *drying zone,* (the area in which water is transferred from the hay to the air), to pass completely through the crop before it starts to deteriorate. If air is very wet, it must therefore be raised in temperature to increase its water-holding capacity. Detailed tables are available for a range of conditions but, for example, if the atmospheric air has a temperature of 60°F (15·6°C) and relative humidity of 90%, an increase of 5°F (2·7°C) in the temperature will raise the water-carrying capacity of an airflow of 1,000 cu ft/min from 18·4 lb to 93·4 lb per 24 hr; an increase of 10°F (5·5°C) will raise it to 171·3 lb. Because of unevenness in airflow, and the resistance of crop to giving up water, especially from bales, the actual gains may not be as spectacular as this. But as a rough guide a temperature rise of 10°F under adverse ambient conditions will give a drying rate equivalent to that expected on a fine summer's day.

There must be a careful balance between the temperature of the atmosphere and of the exhausted air; if the air passing out of the crop has too high a temperature it will be unsaturated, drying will not be very efficient, and the hay at the bottom of the drier may be seriously overdried; if the air emerging is completely saturated and cooled it will deposit water in the top layers of hay. In practice it will generally be possible to maintain the humidity of the exhaust air at 90% during the early stages of drying. But, as shown in Table 7, this corresponds to 30% mc in the crop, so that as the hay dries the rh, which will vary over the crop surface, will decrease to a mean of 80%, or even less.

Control of airflow at the recommended levels and, where heat is used, restriction of heat input so that the exhaust air has a temperature not more than 5°F above ambient, is likely to give the most satisfactory and economical drying.

CHOICE OF SYSTEM

The biggest single factor influencing the choice of system of drying for a particular farm must be the amount of wilting that can be done quickly in the swath, and therefore the amount of water which remains to be removed by artificial ventilation. Drying from high moisture contents, above 55% for example, although technically possible with the fan and heater equipment supplied for barn hay-drying, is inefficient and very costly. Barn-drying should not be used as a substitute for grass drying, which requires properly-designed drying plant operating at a high temperature, and needs a very high standard of management both in the field and at the drier.

Hay for drying will normally be mown earlier than for field-hay, probably when most of the ears have just emerged from the stems and when wilting to 50% or less is not too difficult. Barn-drying of grass cut before ear-emergence is quite difficult, and even after an extended field wilt, several days extra ventilation may be needed to make sure that the hay, which is rich in soluble constituents, does not 'sweat' and deposit water in a layer at the top of the load. Whichever system of drying is chosen, and particularly with bales, which have a much higher bulk density than loose or chopped hay, it is recommended that mean moisture content in store should be reduced to 15%, to ensure that no wet patches act as a nucleus for heating and moulding.

Conditioning of hay, using unheated air, or sometimes air which is slightly warmed, either electrically or by waste heat from an engine or motor, is usually carried out in a *storage* drier. Simple, and often inexpensive, modifications are all that are generally needed to adapt existing buildings for ventilating baled hay. After drying, the hay remains *in situ* until required for feeding, and provided farm layout is suitable little labour is required for feeding out. Alternatively hay may be conditioned in batches in tunnels, in the field, using mobile fan equipment. This hay will generally be moved to final store after completion of drying.

The use of fixed *batch* driers to deal with a few tons of hay at a time, although a very good method of making a high quality product, has declined in popularity because of the double-handling involved. But it remains of special interest to farmers wishing to produce a small amount of very good hay, whilst using silage-making as their main conservation system.

TYPES OF DRIER

For most farmers the preference will in future be for some kind of storage drier. This has a higher capital cost than the small, or the very simple batch drier, but a lower running cost. The overriding consideration is often that bales, once dried, remain in the store until fed. Because hay is loaded progressively, and is stored in depth to be dried over a period of 2-3 weeks, these driers are operated mainly with unheated or slightly-heated air to prevent over-drying, and the moisture content of hay collected from the field should be in the range 35-45%, depending on the type of drier used. This means that hay will usually be wilted in the field for 2 to 5 days, and can be taken from the swath 1 or 2 days before it would be considered fit for baling for normal storage.

WALLED STORAGE DRIERS

Storage driers can be fitted into any building which is suitable for storage of hay, but they should preferably have a height to the eaves of not less than 18 ft and so a Dutch barn is particularly suitable. Existing side-walls may be adequate; they should be reasonably air-tight to within about 1 ft 6 in of the maximum height of loading. There is little lateral pressure from baled hay, so that light walls made from timber supports and corrugated iron or waterproof hardboard can be used, depending on what protection from the weather is needed. The loading door should be wide, preferably not less than 6 ft, as this allows plenty of room for handling, and even enables bales to be lifted mechanically right into the drier.

In some buildings one side has been made of temporary flexible sheeting which can be rolled down to allow easy loading along the whole length of the drying bay. Dimensions of the drier will often be dictated by existing building dimensions, and dry hay requires from 250 to 350 cu ft/ton. Therefore a bay 30 ft x 15 ft with an 18 ft working height will hold from 23 to 32 tons.

The drying platform must be strong enough to take a loading of up to 240 lb per sq ft when the crop is wet. Favoured construction is a flat floor, mounted on blocks raised 2 ft above floor level, and consisting of welded-mesh panels mounted on 3 in x 5 in bearers placed 3 to 4 ft apart (photo 17). To prevent excessive escape of air between the bales and the walls the ventilated part of the floor should either stop about 2 ft short of the walls or be blanked off at the edges with plastic fertiliser bags or similar material.

Rate of airflow through the crop should be between 30 and 45 ft per minute. Use of air-speeds in excess of 45 ft/min can treble fan pressure requirements; for example, at 60 ft/min a back-pressure of 6 in wg (water gauge) could be expected with an 18 ft depth of bales.

Hence a fan suitable for a 30 ft x 15 ft bay should deliver about 18,000 cu ft/min at about 3 in wg. To make the maximum use of this expensive equipment, four or five adjacent drying bays can be connected by a common duct, from which airflow is regulated by control doors; in this way a fan suitable for ventilating 450 sq ft at a time will condition 100 to 150 tons of hay in two months. Alternately if suitable individual air inlets are constructed the fan can be moved from one bay to another and connected to the drying chamber by flexible ducting.

If heat is used under adverse drying conditions, the temperature rise should not exceed 5°F; to obtain this with electrical heating about $1\frac{1}{2}$ Kw is required per 1,000 cu ft/min of airflow.

For this type of storage conditioning hay is best wilted to between 30% and 40% mc before loading. Bales should be stacked with their cut edges downwards to allow air to penetrate more easily. Cross-bonding of alternate layers will control escape of air between the bales and as there are inevitably plenty of small gaps the second and subsequent layers can be packed as tightly together as possible; any very large gaps should be filled with hay from broken bales. Four or five layers may be loaded on the first day and the fan can be switched on as soon as the first layer has been completed. After ventilation for between one and four days the drier can be topped up two or three layers at a time, over a period of three weeks. As many as 16 layers can be loaded, but the moisture content of bales above the 12th layer should be brought down to less than 35%, that is to a level at which hay is frequently baled for stacking in small heaps in the field. After loading has been completed, up to 10 further days ventilation may be necessary, but during the last week the fan can be switched off at night.

UNWALLED STORAGE DRIERS

Many attempts have been made to reduce the capital cost of storage-drying installations. These have included the use of government surplus materials to form the floor and the partial or complete elimination of side-cladding. In the latter case a central tunnel 2 to 4 ft wide and 6 ft high can be erected down the centre of the barn, and the bales, which should not exceed 35% mc, stacked around it. Sufficient air penetrates the stack to control heating, but much of the energy for drying is supplied by oxidation of the sugars in the hay.

An alternative and more reliable method of bale-drying without walls is an adaptation of a Dutch system, used for drying loose or coarsely-chopped hay. A square stack of bales, which can range in area from 15 ft x 15 ft up to 30 ft x 30 ft, is built around a central vertical duct in which is placed an appropriately-sized 'bung' unit. This seals the duct and controls airflow, and it may also contain the fan and motor unit (photo 18).

As the height of the stack is increased the unit is raised so that air is distributed throughout the bales; this is assisted by lateral ducts, which radiate diagonally from the four corners of the central duct in the 1st, 4th, 7th and 10th layers of bales. An airflow of 550 to 600 cu ft/ton of dried hay (10 cu ft/min/bale) is required at a pressure of between 0·5 and 2·5 in wg. In fact, because the area through which the air enters the bales increases as the stack height increases and the distance through which it passes is not increased, back pressure is progressively reduced.

At least six layers should be loaded on the first day, and loading can continue without a break until the stack has been completely built; up to 50 tons of hay (3,000 bales) can be ventilated at one time in the largest practicable unit. After loading has been completed 10-12 days continuous ventilation is required, assuming that the hay has been baled in the optimum moisture content range of 35-40%. Intermittent blowing for a further one to two weeks may also be needed, when drying conditions are favourable, to bring the moisture content down to a safe level of below 20%. The fan is usually installed either on an overhead conveyor or at the end of an underground air duct, so that it can conveniently be used to dry several stacks in a season.

CHOPPED HAY DRYING

An attractive alternative to handling hay in loose or baled form is to chop it so that it will flow as easily as chopped silage and require little or no man-handling. Most types of harvesting equipment used for silage production are suitable for chopping hay. But it is seldom possible to dry hay to a low enough moisture content in the swath for it to be chopped and stored safely, unless it is artificially ventilated. Harvesting should therefore take place at a moisture content of between 30 and 45%, to reduce the time at risk in the swath and to minimise the heavy loss of leaf at harvesting which can be caused by chopping very dry hay.

The chop-length used should be related to moisture content, as both affect density in store and the pressure against which the fan must operate. Moisture content should be less than 40% if a short chop of 2-3 in is used, but may be increased to 45% for a chop length of 4-6 in.

The shorter chop-length produces a suitable compromise between trailer and building capacity (about 350 cu ft/ton), fan pressure, and the ease with which the crop can be distributed within the drier. But at this chop-length a method of air distribution must be used which reduces the depth of hay through which air has to pass to about 5 ft, as the static pressure at 45 fpm is almost twice as high as for bales. Short-chopped hay can also be very dusty, and will compact tightly if blown into store without an efficient means of spreading, so leaving hard spots which heat and mould.

SAFETY MARGIN

Because of variation in moisture content, with some parts of the crop often above 40%, and since spreading is frequently not uniform, a safety margin can be provided by chopping to 4 to 6 in. But this, of course, reduces storage capacity, to about 450 cu ft/ton. An even longer chop, of between 6 and 12 in, has also been used successfully by picking up the hay, which may have been cut with a flail mower to assist wilting, with a flail harvester or double chop machine. Another system uses a continental self-loading wagon fitted with knives to give a random chop, and to improve the load-carrying capacity of the trailer. Such a long chop presents less problems of ventilation, and may allow drying up to a depth of 15 ft or more without the installation of refined methods of air distribution, but it also reduces storage capacity to about 500 cu ft/ton.

The same design of storage drier as that used for bales is suitable, as long as some type of side walling is provided. But even with short-chopped hay about 350 cu ft is required per ton of dried hay of 15% mc compared with 300 cu ft/ton or less for baled hay made under comparable conditions.

For chopped hay-drying a number of special purpose structures are available which aim to make field clearing, loading and feeding a one-man operation. One of these, used successfully on a farm scale for several years, consists of a flat floor storage drier inside a Dutch-barn structure, in which the side walls are moved upwards and out-wards following drying to allow the hay to be easy-fed into mangers placed at the base of the stack (photo 19).

In another barn-based method, developed experimentally at Cockle Park, the building and associated equipment have been designed to give complete evenness of spread of hay within the drier together with a simple method of feeding out. The hay is conveyed from ground level using an elevator, and is then distributed by means

11. Automatic bale sledge at rear forms bales into heaps of 4, which can be handled by the specially designed loader shown in the foreground.

12. Impaler bale loader. Sixteen claws grip 8 bales which can either be conveniently stacked into field heaps of 64 bales, or loaded directly to trailers for haulage.

13. Automatic bale wagon picks up 88 36-in bales which have been dropped singly on edge. Models are available to carry from 46 up to 119 bales, depending on bale size.

14. Tractor-mounted bale bonding unit designed to form unit loads of 20 bales. Manually-formed heaps of 4 are moved mechanically to a loading platform which raises through 90° to form the final heap, prior to hand operation of the bonding device.

15. Front-mounted handling attachment for units of bonded bales. This can be used to form stacks 12 bales (3 units) high, as well as to load and unload trailers.

16. Large rectangular hay bales, weighing between 700 and 1,400 lb have been made and handled successfully, both experimentally and in a prototype field pick-up machine.

16a. Automatic string tying pick-up baler designed to produce bales weighing up to ½ ton.

17. Storage drier floors should be constructed over a plenum chamber of 1 ft 6 in to 2 ft depth and consist of strong timber and 3 in x 3 in square welded mesh.

of a fixed overhead conveyor, a moving cross conveyor and mobile deflector ploughs (photo 20). Apart from reducing power requirement for handling, and eliminating some of the dustiness associated with conveying hay pneumatically, the even distribution allows drying to take place quickly so that up to 5 ft of fresh crop can be added every one to two days.

THE HAY TOWER

Another alternative which has so far had limited use on a farm scale in the UK is the hay tower, which can bring the same degree of mechanisation to haymaking that tower silos bring to silage. Many of these radially-ventilated units are successfully used on the Continent; because they are round, spreading of the hay within the drier, using a permanently installed spreader, is comparatively simple.

One type consists of a circular roof on supporting stanchions, and a deep 'forming' skirt, which takes the place of fixed side-walls. This skirt locates the hay and is lifted upwards as the tower is loaded to form a free-standing stack of 33 ft diameter, in the centre of which is a ventilated duct (photo 21). Height to the eaves is 40 ft when fully loaded and the capacity, clearly depending on chop-length and loading moisture-content, is between 120 and 150 tons. Another type of tower, much higher in capital cost per ton stored, has permanent side-walls made of perforated steel panelling. Both are unloaded mechanically down the centre ventilation duct, so that man-handling can be completely eliminated.

A disadvantage of chopping hay for handling is that the weight of trailer loads is low, with a crop density of only 3 lb/cu ft, while the system employed usually has no break-points, to ensure a quick turn-round and an adequate supply of trailers to keep the harvester at work. Therefore, although output can be good in terms of man-hour/ton the total rate of handling is likely to be low, at perhaps 10 to 15 tons a day.

However, chopped hay-drying remains an attractive proposition for farmers who wish to make up to 150 tons of very high-quality hay, possibly in conjunction with silage, who are prepared to wilt to less than 40% mc and to continue hay making over a period of three or four weeks.

*　　*　　*　　*

Many farmers will continue to make hay in spite of the difficulties because they consider that it is a good and predictable type of feed for their stock and, especially when barn hay-drying is used, that it

D

will give them the production they require from grassland. Reference has also been made in Chapter 2 to a possible increase in the use of chemical additives, such as propionic acid, which for about the same cost as barn hay-drying may reduce swath exposure-time, and above all control heating and moulding in store. But one of the severe limitations of most hay-making methods at present in common use is that, because of extreme weather dependence, there is no guarantee that the crop can be harvested at its optimum stage of growth, particularly if mowing must be geared to the throughput of a barn hay-drier.

One of the most important reasons for the present increase in silage-making at the expense of hay must be that the handling and storage methods employed allow the bulk of the grass crop to be cut and stored within 10 to 14 days. Particularly where no wilting is involved the conservation programme can then be planned and carried out to schedule in at least four years out of five.

Chapter 7

SILAGE MAKING

INEVITABLY THE day will come when the silage-maker faces the fact that his consideration of techniques is at an end, and he must get down to doing the job. He should have weighed up such factors as the type of crop he will grow, the animals he wishes to feed, and the farming layout within which he must work, so that certain practical priorities will be clear in his mind.

These will be influenced by the theoretical considerations we have outlined, but it is undeniable that many a farmer finds it difficult to plan and organise a silage-making programme in the light of the conflicting exhortations of his various advisers.

The following six key priorities will assist in clearing the mind and arriving at workable conclusions:

(1) Cut the crop at the optimum stage of growth.

(2) Achieve, if possible, a minimum crop dry-matter content of 22%.

(3) Ensure that the crop coming into store is chopped to the correct length.

(4) Eliminate the chance of soil contamination at any stage.

(5) Fill the silo by a method that prevents air movement, heating and oxidation.

(6) Completely seal the silo against entry of air as soon as filling has been completed.

STAGE OF GROWTH AT CUTTING

Enough has already been said to underline the importance of this factor, and the establishment of the characteristic D-value curves for most varieties of crop has been appreciated by many grassland enthusiasts; yet failure to comply in practice with this recommendation is still widespread. The common temptation to wait a few days for a 'little more bulk' will seriously affect crop quality. The fact that the stage of growth of the crop at cutting will have more influence on the eventual feeding value of the product than any of the other factors under the farmer's control accounts for the high priority we give to it.

DRY-MATTER CONTENT

Increasing the dry-matter content of silage is now regarded as a desirable end in itself, and in many instances is of major importance in the farmer's strategy. We have already discussed the role which dry-matter has to play in increasing the concentration of nutrients in the grass, and some authorities have concluded that a high-dry matter *must* be achieved at all costs before ensiling the crop. But the D-value curves give a clear indication of the optimum stage at which each crop should be cut, and obviously this time is of short duration. It is when considering these two conflicting problems, of the need to cut at the right level of D-value, and of the desirability of high dry-matter content, that the impact of weather conditions on a wilting programme should be taken into account. In a climate where regular rainfall ensures ideal conditions for grass growth it must be accepted as unlikely that we shall get equally ideal conditions for high-dry matter conservation.

Local weather records provide the monthly distribution of rainfall, but of more importance are the likely hours of sunshine, since we are interested in the incidence of drying conditions rather than in the absence of rain. Much of the grass cut for conservation in Britain will be harvested during May and early June, and weather records during this period make clear how difficult wilting can be. This is also the time when the moisture content of crops is at its highest. Although it may be possible to increase the dry-matter content of a sappy crop on the day of cutting by applying a severe conditioning treatment, two consecutive fine days are needed to increase DM to the range 25-30% (Table 6), and then only when mechanical treatment is given immediately after the swath is mown.

The main reason for selecting silage-making in clamps or in bunkers as the preferred means of crop convervation is to achieve as great a degree of independence of a fine weather requirement as possible, and for this reason the farmer will frequently accept a compromise between the weather and the crop. This is embodied in the 'minimum wilt' concept, in which cutting takes place at the precise time indicated by the D-value curve, while a target of dry-matter at the lower end of the scale, i.e. about 22%, is accepted. At this level of wilting there is a useful reduction in the weight of crop that has to be handled, little effluent will flow from the silo, and the time that the crop is at risk to the weather in the swath can be reduced to 24 hours or less.

There *will* be occasions when favourable weather coincides with harvesting, and the level of dry-matter we are suggesting will be easily achieved, but under these conditions it is necessary to keep tight control to avoid over-drying in the swath. By adjusting the interval between using the cutting machine and the harvester a good measure of control is possible, though varying weather conditions and overnight dews make this quite a difficult calculation. Cut forage above 30% DM can lead to problems in the silo caused by the retention of air within the crop, and the failure to expel this air can cause heating and mould formation. To some extent this can be controlled by the length of chop, and this brings our next consideration into focus.

CROP SPECIFICATION AND HARVESTERS

It is vitally important to bring the same degree of precision to the way the crop is handled that we have already suggested must be applied to the timing of the harvesting operation. The degree of chopping of the crop will depend on the type of silo to be filled; just as with tower silos a chop-length of 1 in is a maximum, so can similar criteria be used in the case of surface silos. These standards directly relate length of cut to the target DM, and this table gives a guide.

Target DM	Maximum chop-length (in)
under 20%	8
20%—25%	5
25%—30%	3
above 30%	1

This relationship between the storage system, the dry-matter content and the length of chop required will largely determine the type of harvester to be used, and how it is to fit into the silage system selected.

HARVESTERS FOR SILAGE-MAKING

The flail harvester has the advantages of simplicity and low capital cost. Several makes* can easily be converted to do any one of the following three operations:

(a) to cut and harvest the crop directly;

(b) to cut the crop and leave it in the swath in a partially chopped and lacerated state for wilting; and

(c) to pick up the wilted crop from the swath.

This flexibility can be most useful where the farmer may wish to wilt, but does not want to duplicate machines.

Amongst the drawbacks of the simple flail harvester is the tendency to exert a powerful aspiration effect on the crop and adjacent

* These are discussed in MAFF Mechanisation Leaflet No. 13.

ground areas, which can suck soil up with the forage. This may lead to serious contamination which can upset subsequent fermentation. This undesirable result can be aggravated if the machine is set to cut so close to the ground that the spade-shaped flails scalp the sward and pick up earth and small stones. The main disadvantage, however, is that most of these machines cannot be adjusted to chop the crop shorter than 6 in, and will pass a lot of material 10 or 12 in long. This limits the amount of crop that can be loaded into a trailer (Table 8); it can also lead to a considerable volume of air being retained within the forage, causing poor consolidation in the silo and slow establishment of anaerobic conditions. Self-feeding of animals is difficult, as intake is restricted because the material is firmly retained in the mass of silage. The long straggly lacerated herbage is also not easy to handle with normal mechanical unloading equipment.

The double-chop harvester (photo 22) is more complicated than the simple flail. It can perform the same three functions, mowing, mowing and loading, or picking up the wilted crop, although the chop-length produced by the edge-cutting curved flails, when it is used as a mower, may be too short to allow efficient pick-up and loading after wilting. So, if this harvester is to be used for picking up from the swath, the best results will be obtained by mowing with a machine designed for the job; this will leave forage of 8-10 in length, set up in a swath of a satisfactory shape for wilting and loading.

For many farmers, however, this type of harvester which can cut and harvest direct as well as pick up cut crop from a swath is most useful, for a convenient and effective working procedure is to load the crop mown on the previous day until the middle of the afternoon, by which time the standing crop is surface-dry and can be cut direct; this will reduce the work-load on the pre-cutting machine and can avoid the harvesting of over-wilted crop. It also permits harvesting to continue uninterrupted if the pre-cutting machine breaks down.

The double-chop harvester produces a fairly uniform material in the range 3-6 in which considerably increases the weight of crop that can be carried in the trailer; the resulting forage is well-suited for efficient filling and settlement in surface silos, for self-feeding, and for loading out. The design and arrangement of the cutting flails result in minimum suction being exerted on the crop, and there is less risk of soil contamination than with the normal flail machine.

Several points in favour of short-chopped material lead to the conclusion that the precision-chop (metered-chop) harvester is the best machine for use with surface silos (photo 23). The higher initial cost is compensated by a big increase in potential output, which can be of crucial importance to the large-scale operator—although this higher output will not be achieved unless the machine is matched by commensurately high capacity in both the pre-cutting and hauling elements of the harvesting team. The drive shaft and gear box of the precision-chop harvester are capable of transmitting considerable power to the cutting mechanism, and this fact —resulting in higher output—demands a powerful operating tractor.

The shorter chop, $\frac{3}{4}$-2 in, allows much more dry-matter to be carried in each trailer (Table 8), an important aspect where long hauls are involved, and the chopped material is also easy to handle with loading machinery at the silo. Consolidation caused by the inert weight of the cut crop is very rapid, and gives excellent control of air movement and heating. The silage can be easily self-fed, or loaded by front-end loaders, and will flow more readily than long material in mechanised feeding systems.

Because precision-chop machines are used mainly for lifting wilted herbage, they are normally equipped with a pick-up attachment. On stony land the pick-up tines tend to lift stones with the crop, and these can cause damage to the precision-chopping cylinder. This is particularly liable to occur where stones have been moved into the swath in the process of side-raking two swaths into one, and serious delays can result. So common is this problem on some soils that harvester operation has to be limited to the single width of the pre-cutting machine, with a considerable reduction of output potential. Most models can be fitted with a cutting head, usually based on a reciprocating cutter-bar, although units with flail-cutting cylinders and rotary discs are also available. However, these attachments add considerably to the cost and power requirement of the harvester, as well as taking time to fit and to remove.

Precision-chop machines are essential where the crop is to be loaded into a tower silo, since the unloading machinery will not deal efficiently with the longer unevenly-chopped material from other types of harvester. There are also particular crops, such as maize, which can only be satisfactorily handled by this type of machine.

TRAILERS WITH DIFFERENT TYPES OF HARVESTER

Single-cut flail harvesters are of two types, those designed to fill a trailer towed integrally behind the harvester, and those which will in addition load a separately-towed trailer running alongside. The former has the advantage that a second tractor and operator are not required; further, although there is often a higher output per hour from side-loading, because the job of coupling and uncoupling the trailer is avoided, the output of in-line systems can be increased by using a hydraulic pick-up hitch. However, on sloping fields and soft land there is considerable advantage in the separately-towed trailer, since the tractor-harvester-trailer train can get into trouble through wheel slip, bogging down or overturning.

Some flail harvesters are mounted on the side of the tractor and elevate the crop into a trailer drawn directly behind it. They are particularly suited to operation by one man, as the harvester can be quickly uncoupled, leaving the tractor and trailer free to haul crop to the silo.

Most double-chop and precision-chop harvesters will fill either a trailer towed behind them or one running alongside, but when high-dry-matter crops are being loaded care must be taken to screen the top of the trailer so that strong winds do not blow away the light and leafy particles.

Another major cause of delay in the transport system arises because badly-designed trailer doors and fittings lead to jamming, so that brute force must be used to open doors to remove the crop. Even when the doors *are* open, the compacted crop can only be emptied out by jerking the trailer forward on the tractor clutch. The most effective type of rear door is the up-and-over counterbalanced gate, secured by a spring latch at the base; certainly the ease of opening under load should be one of the main considerations when a new trailer is purchased. For easy emptying there is great advantage in having tapered trailer sides, wider apart at the rear than at the front; standard trailers can be modified quite easily by fitting false hardboard sides within the fixed sides. The hardboard is supported on wood studding, with the smooth side facing inwards, so that the internal width of the trailer is narrower at the front than at the rear. When the trailer is tipped the load slips out easily from between these flared walls (photo 24) and the time saved more than compensates for the slight reduction in the amount of crop carried.

SOIL CONTAMINATION

The level of soil contamination in cut forage will vary with the type and design of the harvester used, and may be increased by mud-splash during heavy rain. More than 10% of the crop dry-matter must be soil before the resulting silage looks and feels 'dirty'; this can be compared to the figure of 11% or more which dairy cows may take in with the grass they graze in September. Pit silage made under bad conditions with grass buckraked directly from the field can contain over 20% of soil; but in grass cut by flail harvesters *correctly adjusted* there may be little more soil than in samples cut by hand.

This emphasises the importance of correct setting of cutting equipment, and particularly of the height of the flails above the ground surface. They must not be allowed to clip the surface, either when cutting direct or when lifting from the swath; as noted earlier, setting the height will be easier if swards are heavily rolled in the spring. Soil contamination can also occur from the tyres of tractors buck-raking crop into the silo; the provision of a concrete loading area with a 4 ft retaining wall at one side is a most worthwhile investment, and with outdoor silos this will markedly improve self-feeding and tractor unloading of the silage in the winter.

Filling the silo: The Dorset wedge method

Walled silos

The subjects of stage of growth at cutting, dry-matter and crop specification have, so far, been considered in abstract form. But there is often scope for much heart-searching on the day the decision is taken to start harvesting. Simple matters left undone until that day, such as the overhaul of harvesters, the preparation of trailers, and a check of their unloading gear, are the cause of quite unnecessary delay. It is remarkable how frequently farmers leave silo preparation and the attempt to purchase polythene sheets until the very morning cutting is due to start. Although the start may sometimes have to be put off for a day or so because of really bad weather, far more often cutting can begin on the date intended. With all the equipment and the silo fully prepared during previous weeks the whole sequence of cutting and harvesting can swing into action as planned.

Polythene sheets, used to control wastage on the wall surfaces of silos, should be attached to the silo before filling commences. Except where the walls are airtight (rendered concrete or resin-bonded plywood facing), and certainly in all sleeper-walled silos, the side-sheets should come down to ground level to prevent air entering.

CONTROLLING SHOULDER WASTAGE

Side-sheets to control shoulder wastage are fixed to smooth walls as shown in photo 25. The wall is coated with bitumen paint (National Coal Board: black bitumastic) applied with a long-handled soft brush. After about 15 minutes this will become sticky, and the sheet of polythene can be applied, rather like hanging wallpaper: even better adhesion is obtained if the wall is already covered with a coat of bitumen paint applied some weeks earlier. 300-gauge polythene sheet is preferred, as any thicker sheets are too stiff and heavy, and tend to peel off the wall before the bitumen is dry. For most purposes a 6 ft wide sheet of polythene is placed lengthwise along the walls with half the width of the sheet stuck to the top 3 ft of the wall and the remaining 3 ft tucked away on top of the wall.

This is pulled across the 'shoulder' when the silo is finally filled, a coat of mastic put on the upper surface and the top sheet then put in position and sealed by pressing the two sheets together (fig. 7).

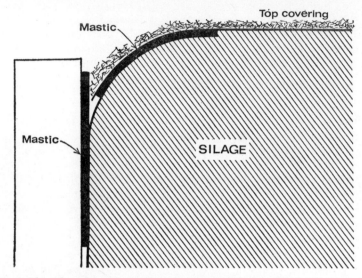

Fig. 7. The side and top sheets prevent wastage on the 'shoulders' of bunker silos.

As the silage settles, the weight of covering material on top of the sheet makes it peel away from the upper part of the wall: but the silage still presses the bottom 2 ft or so of the side-sheet against the wall, so that no air can enter. With this method the wastage commonly found at the 'shoulders' of silage made in a walled silo can be almost completely prevented. The same procedure obtains for end walls, including those put in position before filling commences and designed to be removed to give access to the silage when feeding out.

AVOIDING OXIDATION AND HEATING

The first loads that come in can be dumped directly against the back wall of the silo, and packed close beside each other. Since the aim is to avoid oxidation and heating, the first consideration is to achieve a deep daily fill to assist consolidation and to act as a 'blanket' on previously-filled material. A fill of approximately 3 ft of cut forage will have these effects. In the case of silos of small dimensions it may be possible to fill to 3 ft over the whole area of the silo floor on the first day, but in most cases the silo will be much too large to allow this. The need to ensure the minimum depth of fill specified above will then determine the area covered. The most effective way is to fill the silo in the form of a wedge, with a sloping face up which the loading tractor runs.

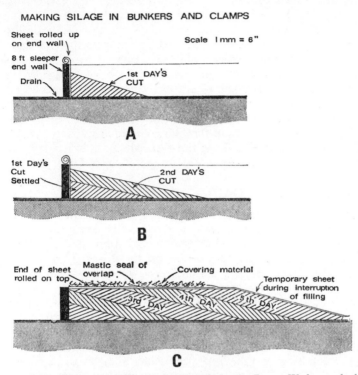

MAKING SILAGE IN BUNKERS AND CLAMPS

Fig. 8 (A-C). Stages in filling a bunker silo by the Dorset Wedge method.

The first day's fill will have the wedge shape shown in fig. 8A. The slope established will depend on the width of the silo, related to the speed with which the crop is coming into store, and in the majority of cases will approximate to the 20° illustrated. Slower filling may require a steeper slope, as will the filling of a wider silo; but the slope must also be shallow enough for the tractor to run up and down with complete safety. As soon as the last load is put in position the exposed surface of the crop is covered with polythene sheet to prevent air movement and heating.

The second and subsequent days' filling will be added to the silo, as shown in figs. 8B and 8C, in daily layers of not less than 3 ft of material, still maintaining the original slope, and progressing towards the open end of the silo. When the end is reached the grass is levelled off to give a uniform finish, so that the whole of the material in the silo is at the desired depth. Filling can be carried out by a variety of equipment but it is important to adhere to the procedure just described, which ensures that any one section of the silo is filled from floor to finished height *within three days.*

The main advantage of this method of filling is the flexibility in organisation that it allows—and it should not be thought that the standards indicated here can only be achieved by a large labour force and lots of machines. In practice the Dorset Wedge system is successfully used on farms where *one* man does the whole job of cutting and filling. Equally it has proved very effective in the hands of the large operators in the country, including contractors. The key to success is in variations in the filling slope, in relation to the harvesting rate and the width of the silo, which allow control of the fermentation process under a wide range of conditions.

LOADING INTO SILO

The cut crop is usually loaded into the silo by a tractor and buckrake. The preferred type is rear-mounted, as driving forwards up the slope before the crop is well consolidated can be difficult—although looking backwards can be a strain on the driver. It is sometimes difficult to get the crop off the buckrake when the tractor is still on the slope and a much more positive discharge can be obtained by using a push-off ram on the buckrake. Outputs up to 30 tons/hour are possible although some skilled operators consider that the time taken to operate the ram can reduce their overall loading rate into the silo! Alternatively a high-lift front-mounted buckrake can be used (photo 26). Use of a high-lift foreloader reduces the problem of front wheels sinking into the crop, and high rates, up to 35 tons/hour, can be obtained with a foreloader fitted with a hydraulically-operated grapple fork (photo 27). The increased lift and discharge height allows accurate placing of the loads of grass, but gives less immediate consolidation than when a buckrake is used. A similar, but more severe problem arises when a dump box and blower are used to load crop into a bunker silo, because the consolidating effect of the load bouncing in the transport trailer is lost. Much care must then be taken in covering the surface each night to prevent air movement through the loosely-packed crop.

Whichever method of loading is adopted there is an advantage in filling the silo progressively on a wedge, as this allows the second element of the Dorset wedge method to be carried out—namely, the final covering and sealing of the silage surface with plastic sheeting as soon as the finished height is reached. The aim is to fill each section to about 1 ft above final settled height within three days, and final covering progresses from one end of the silo to the other as filling proceeds (fig. 8C). As already noted, the exposed slope should always be covered with sheeting during breaks in filling, both overnight and at weekends, with a slightly-weighted polythene sheet that can be easily removed.

CONSOLIDATION

The passage of the buckrake tractor as it loads material into the silo will result in some consolidation, but if the chop-length of the crop bears the right relationship to its DM, the weight of the crop itself will result in a considerable degree of settlement. Some rolling may be necessary during filling, but in general farmers spend far too much time on this job, and there is evidence that the constant passage of a tractor over the surface can produce a 'bellows' effect, which actually assists the entry of air. On no account should rolling be allowed to delay the final covering with the polythene sheet, since heating and wastage are certain to develop if sealing is not rapidly completed.

As soon as a section of the silo is filled the appropriate length of side-sheeting is pulled over the 'shoulder', mastic is applied, and the top sheet is laid down and pressed into contact with it (photo 28). It is then most important to apply some weighty material over the whole surface of this sheet so that it is immediately pressed into tight contact with the silage below to prevent entry of air. A 4 in layer of chopped grass will do this, and is particularly recommended if straw bales are to be added later, since the grass will prevent the sharp ends of the straw from perforating the sheet. Other suitable materials are farmyard manure (particularly for outdoor silos), pit-belting, and old tyres if these are used in sufficient numbers. The whole area of the sheet *must be completely covered from view;* this is particularly the case with outdoor silos, where any exposed area will be weakened by sunlight and perforated by the claws of birds.

HIGH MOWTHORPE FILLING PROCEDURE

A method similar to the Dorset Wedge filling procedure has been used at the Ministry of Agriculture's Experimental Husbandry Farm

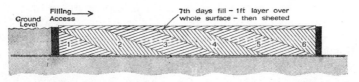

Fig. 9. Filling a bunker silo installed in sloping ground.

at High Mowthorpe, where loading has to take place over one end wall of the silo due to the existing ground levels (fig. 9). In this case the sloping surfaces are maintained in the way described earlier, but since material comes in from the highest point the final covering sheet cannot be applied until the whole of the silo is filled. This is compensated for by applying a 12 in layer of unwilted material over the whole top level area as the final operation immediately before covering with polythene sheeting and weighting this down.

OUTDOOR SILOS

So far we have been concerned with covered walled silos under Dutch barns, for which 300-gauge sheeting is both economical and effective. It is, however, a reasonable assumption that if a satisfactory job can be done in protecting silage indoors from the ill-effects of air movement, then the same methods can be used to give weather protection as well for outdoor unroofed silos. The main difference is that polythene sheet of a minimum of 500-gauge thickness, or butyl rubber sheeting, must be used.

Much the same filling and covering procedure will be used for outdoor *walled* silos as for indoor ones, but there is a special need to give a weatherproof finish and to avoid the ill-effects of rain running down the insides of the walls into the made silage.

OUTDOOR WALLED SILO-END VIEW

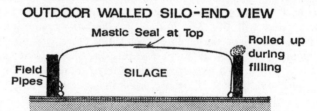

Fig. 10. Sealing an outdoor bunker silo to prevent entry of rain-water.

Fig. 10 shows an effective method of using polythene sheets on each sidewall; when filling is completed both sheets are drawn across the silage to overlap at the highest point, where a bitumen seal keeps out air and rain. Any water on the outer surface of the sheet finds its way to the top of the walls, where it can disperse between the sheet and the wall and be carried away by the field pipes at the foot of the walls. This is also a very effective way of sealing porous walls of any type. As with all other surface silos a gently sloping site and good drainage are particularly important.

Clamp silos

The ultimate in economy and flexibility for storage is the outdoor silo without walls, since this can be placed at any convenient point on the farm where there is a sloping site on clear ground. A concrete base will be needed, of course, if self-feeding is to take place, but in other cases quite adequate results will be achieved on a normal field surface, except where the soil is very heavy.

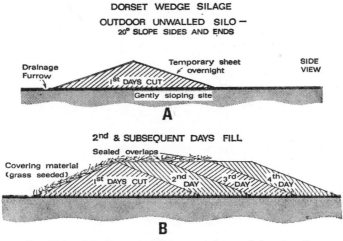

Fig. 11A and B. Stages in filling an outdoor sealed wedge silo.

Filling these silos follows exactly the same principles already described for indoor silos, but some simple adaptation is necessary to compensate for the absence of walls, and to ensure that the covering sheet is pressed firmly against the silage so as to prevent wastage. The silo sides should have a slope no steeper than 20°; the first day's stage (fig. 11A) shows the end slope and the filling slope established. The second and subsequent days' filling (fig. 11B) proceed with the addition of a minimum of 3 ft depth of crop per day, maintaining the shallow slope on the sides. As soon as the silo is long enough, and at most after two days' filling, a polythene sheet (500g) is applied laterally across the silo (photo 29) and the filling tractor places a layer of covering material (farmyard manure, soil, sawdust etc) on top (photo 30). Further applications can be made later from the side of the silo by a rotary spreader throwing material on to the sheet. It cannot be too strongly emphasised that this must entirely cover the whole surface of the sheet so that at no point is it exposed. The covering layer should also be at least 6 in deep to allow for the washing effects of rainfall: it is here of course that the shallow slope to the sides is so vital.

Another useful measure that will help to anchor the top material on the sheet is to make a light sowing of grass or cereal seeds. These will root into the farmyard manure (or soil) and bind the whole together against the effect of rain; they will also make removal of the covering a great deal easier, since the whole mass will be rooted and can be pulled off in a mat.

ELIMINATING SURFACE WASTAGE

Prompt sheeting and cover weighting will go a long way to eliminating surface wastage, even in the case of sloping sides that have not been consolidated in any way. Any delay in sheeting and sealing will result in heating and oxidation of nutrients and in surface wastage which will render the whole effort a waste of time.

Great care *must* be taken by the tractor driver in building this type of outdoor clamp. But the risk of overturning is much less than when the conventional type of clamp with vertical walls is being made, as it is not necessary to consolidate the sloping sides to prevent overheating if they are properly sealed.

This outdoor clamp method was succcessfully demonstrated at the National Agricultural Centre in a silo filled on May 24th/25th, 1971, which was shown to be entirely free from wastage when it was opened for exhibition at the Royal Show in July 1971.

Silage additives

The role of chemical additives for silage was discussed in Chapter 2. The main advance has been in the development of effective applicators, rather than in the chemicals themselves, many of which were tested years ago. Applicators for chemicals in both solid and liquid form are available, with liquid applicators the most effective because they are less liable to block in wet weather. Where only a few hundred tons of silage are to be made equipment based on a 5-gallon plastic container of chemical feeding directly through a tube into the flail or chopping mechanism is satisfactory (photo 1). It is important to insert the feed pipe at a point on the harvester where air suction will act upon the aperture through which the nozzle passes. It is not uncommon to find that the manufacturers of both harvesters and applicators have failed to identify these positions correctly, and the fitter should experiment with a simple smoke generator to get the best location. Back pressure will cause fumes to escape, and may give rise to a damaging spray that will harm the eyes and skin. Where larger amounts of silage are made some operators have fixed a bracket to the harvester to carry a 40-gallon drum of additive, this being fed into the cutting mechanism either by gravity, or preferably by a simple glandless pump.

The amount of each particular additive applied should be as

18. Vertical central ventilating ducts for drying by the Dutch system are controlled by a "bung-unit", which may have the fan mounted in the top of it, if roof bearers can carry the weight. An underground, or surface duct within the heap of bales can also be used.

19. Cows feeding at the base of a chopped hay storage drier. One man can feed 100 cows in from 10-15 min.

20. Conveyor distributor installed in chopped hay drier at Cockle Park.

21. Tower hay-drier consists of a raised slatted ventilating floor and a steel structure which raises a roof and hardboard former as the stack is loaded by blower and levelling rakes. Hay is unloaded by a series of angled finger wheels which feed it out through a side door.

22. Double-chop forage harvesters mow the crop with a series of edge-cutting curved flails and feed it via an auger to a flywheel-type chopper thrower. Chop length, although very variable, is mostly from 2 to 6 in.

23. Grass can be mown directly, but is generally picked up from a wilted mown swath, with a metered-chop forage harvester. Both cylinder and flywheel-type choppers are available.

24. Trailer with flared sides, wider at the rear than at the front, and with automatically-operated lifting tail-gates ensures rapid emptying at the silo site.

recommended by the manufacturer. This will depend partly on the crop (lucerne needs a higher level than ryegrass) but mainly on its moisture content, with wet crops needing more additive. Rate of addition is generally controlled by the size of the nozzle at the end of the delivery tube leading into the cutting mechanism, or by the speed of rotation of the pump, where this is used.

Some additives, particularly those containing acid, are corrosive and can damage eyes and skin. In handling them special care must be taken to follow the manufacturer's instructions, particularly with regard to the use of gloves and goggles during filling and changing of containers. It is also important to cut off the flow of chemical whenever crop is not passing through the harvester, as on the headlands of fields, so as to avoid spray drift. This is generally done manually, but the wider use of automatic cut-off devices would simplify this operation.

Accurate and uniform application of the liquid to the crop is of crucial importance to this whole procedure. Thus it is a real advantage to explain to the harvester operator why the additive is being used and how it works. He is then much more likely to apply the correct rate, and to reduce this at intervals during the day as the crop dries out—or to increase the rate if rain falls. If possible a number of trailer-loads of crop should also be weighed, for few farmers really know how much their trailers hold—and unless they do, the correct nozzle size to use is a matter of guesswork.

MINIMUM FERMENTED SILAGE

At the time of writing this whole 'new' subject of silage additives was only just opening up, and results were not available which would allow different commercial additives to be evaluated. As was noted in Chapter 2, formic acid has been shown both to improve silage fermentation and to increase level of voluntary intake, but there is much less information on additives, based on formaldehyde, which aim to restrict fermentation. It seems certain that further research in this field will improve our control of the ensilage process under practical farm conditions.

Although additives are likely to produce entirely new types of silages for storage feeding, the procedure for making them is similar in almost every way to the method we have outlined for ordinary silage. This similarity is in itself a potential impediment to the fullest exploitation of these feeds, since farmers may be inclined to treat them as simply a modification of the conventional material, and it may well be desirable that a completely new name should distinguish the products.

Minimum fermentation procedures can be applied to both un-wilted and wilted silage. The greatest benefit is likely to be with direct-cut or slightly-wilted crops, which can be the most difficult to ensile; in this way the losses often associated with heavy wilting can be avoided. However, additives may also be useful in preventing overheating in high dry-matter wilted crops, which is often difficult to stop by consolidation and sealing alone.

Precisely the same method of quick filling in regular daily incre-ments, and equally prompt application of the polythene sheeting will be followed for these new types of silage as has already been described.

Care during storage may be even more important than with conventional silage, since the lower content of fermentation acids may make the silage more vulnerable to spoiling organisms and oxidation if air is allowed access.

A noticeable feature of the products, particularly those based on formaldehyde, is the absence of a typical 'silage' smell and this will commend it to many people. There is also some evidence that the fact that less fermentation takes place results in a reduction in the volume of effluent production.

Of principal importance, though, is that additives are likely to ensure a product of higher nutritive potential, and advantage must be taken of this if the cost of the additive is to be justified. The way the farmer approaches the planning of his crop conservation may well be modified, by this fact, to the view that it is well worth taking considerably more care over the process of making, and of feeding the resulting silage.

EFFLUENT LOSS FROM CLAMP AND BUNKER SILOS

Compacted wet crops lose moisture, which runs out of the silo as effluent. The loss of dry matter as effluent is seldom more than 1% of the crop, but it is a valuable fraction, containing readily-digested soluble components and minerals. More seriously, effluent creates the social problem of smell, and the legal problem of pollution of water-courses. So every effort must be made to limit effluent—and where it *is* produced, to trap it so that it does the minimum of damage.

Little liquid flows from crops of over 25% DM in a bunker silo, over 35% in the lower half or over 30% in the upper half of a tower silo. This coincides with the general advice to wilt the crop whenever possible—still the most effective way of reducing effluent. But under adverse weather conditions it is often necessary to load crop at under 20% DM, at which level water can be squeezed from it by hand. The main measure then is to avoid heavy consolidation. Heating must be prevented by careful sealing, for heating and oxidation of sugars produce water and can further increase the moisture content of the crop. The effect of additives on effluent flow is not yet clear; formic acid seems to speed up the first flow, but may produce no more effluent than the control silage, while formaldehyde may reduce effluent.

Where effluent is produced it must be trapped by half-round drains laid across the fall of the surface on which the silage is made, and led away to a catch-pit. From this it should be diluted and sprayed out on to the fields where the silage was cut; if this is done every day or so, before the effluent begins to decompose badly, a few hundred gallons per acre will not damage the sward.

VACUUM SILAGE

In the early 1960s the technique of vacuum silage was introduced to the UK from New Zealand. The aim of this process is to produce an anaerobic state within the silo by exhausting the air remaining in the crop with a vacuum pump. The removal of air limits oxidation to a minimum, though it appears unlikely with crops below about 25% DM that a significant volume of air will be trapped within the silo after the initial settlement has taken place. However, the main effect of the vacuum silage process is the very considerable compression that is exerted on the outside of the polythene envelope by atmospheric pressure. Particularly in high dry matter crops, i.e. over 25% DM, this can have real benefit in circumstances where consolidation by other means is difficult because the material is light and spongy.

The silage must be contained in a complete polythene envelope to obtain air-tight conditions. A ground-sheet is placed in position and the first layers of grass placed upon it. Damage can be minimised by rolling most of the ground-sheet on a length of rod or water piping so that unloading can take place over the rolled-up sheet. The remaining sheet is progressively unrolled to accommodate the base layer until the whole of the area required is covered.

Filling proceeds as for the other surface silos we have described, and should be completed as rapidly as possible. The top sheet (500g minimum thickness) is then laid over the stack of crop and sealed to the base sheet with a proprietory sealing-strip of concentric tubing which grips the sheet edges together. A vacuum pump is attached to the silo by rigid piping that will not collapse, and the pump is operated until all the air is removed. With large silos a high-capacity vacuum pump removing up to 100 cu ft/min at 15 in mercury is required to exhaust the air satisfactorily, but for smaller clamps milking machine and vacuum-tanker pumps with a capacity of about 40 cu ft/min at 10 in mercury can be effective. It is, however, important to maintain a vacuum until the pack is opened to load more crop, but this can prove difficult. Having removed the air it is essential during long-term storage to keep the polythene sheets in close contact with the silage by covering with some weighty material over the whole of the outside surface area, as has been described with Dorset wedge silage. Failure to do this accounts for many of the unsuccessful results from vacuum silage. For surface wastage occurs wherever the sheet is not kept in contact with the silage beneath.

CARE DURING STORAGE

Between the sealing of a silo and opening it for feeding six months will usually elapse, and during this period occasional examination will be well worthwhile. Wind and rain may cause the sealing sheet to be exposed, and these areas should be covered immediately, to keep the sheet in contact with the silage beneath.

Outdoor clamps should be fenced off, since it is common to find they have been walked on by cattle, sheep or even deer, with disastrous results to the sheets—and to the silage. The silo should also be checked regularly to see that no rain is getting in, and that there is no faulty drainage of the site, causing the lower layers of the silage to become water-logged.

Provision for sampling

At the time of covering the silo provision should be made for sampling the clamp by marking with a piece of distinctively-coloured material such as polythene fertiliser sack at two or three suitable points where the sheets overlap. This will avoid breaking the air-seal more than is absolutely necessary when a core-sampler is used. The Agricultural Development and Advisory Service will sample

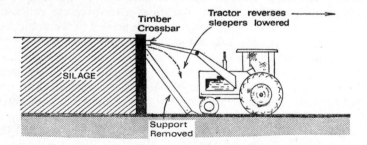

Fig. 12. Use of a tractor fore-loader to hold a silo end-wall while the supports are removed.

all types of silage, and the regional nutrition departments provide the farmer with analytical details and feeding recommendations designed for the stock to which the silage is to be fed.

Opening the silo

Considerable care is necessary when removing walls that retain the silage in order to expose the material for feeding. Silage exerts a powerful side-pressure, particularly when filling goes above 6 ft in height, and where heavy consolidation has taken place. In the case of temporary sleeper walls a tractor fore-end loader fitted with a timber crossbar can support the sleepers whilst other supports are dismantled, and careful backing-up and lowering (fig. 12) can avoid a dangerous collapse. As the external walls are taken away the polythene lining sheet should be revealed firmly adhering to the silage, and when this is cut away the silage surface should be completely free of surface wastage (if the making and storing suggestions described here have been followed successfully). The days when the farmer spent a day hauling away rotted material before he reached the edible part of the silage must be a thing of the past with both indoor and outdoor silos.

Some of the methods we have described for making clamp and bunker silage may seem elementary, but their aim is to produce an entirely predictable material free from surface wastage. Particularly where high dry-matter silage is the target, adverse weather at harvesting can lead to serious loss and disappointment, and we suggest that the method chosen should eliminate, as far as possible, factors that are outside the farmer's control.

It is only necessary to make the most cursory check on current silage-making methods to see how widespread surface wastage remains. It represents one of the most serious misuses and depletions of expended resources in the whole range of agricultural activities, and in many cases only a minor change in methods and their timing can promote a dramatic improvement in results. This is within the grasp of everyone making silage.

The more sophisticated and costly harvesting and handling procedures will need to be examined closely to ensure that they contribute a specific economic advantage. In particular where a higher-value product is the result it is necessary to calculate at what cost this has been achieved, and to compare the result with the cost at which alternative purchased feeds could produce the same production response. As the price of these feeds rises the justification for more expensive and efficient conservation procedures will become more valid.

Towers for silage

We have paid particular attention to the filling of surface silos because we think that most silage-makers will, and should, adopt this method of storage, at any rate for the foreseeable future. Many British farmers, however, have seen tower silos in the United States, where they are especially valuable for the ensiling of lucerne and whole crop maize (corn), as part of a fully-mechanised system of livestock feeding. As a result a considerable number of storage towers have been installed in the UK, in an attempt to capitalise on the efficiency which a sealed container and full mechanisation can contribute. This development has tended to be localised in certain areas, such as the south-west coastal area of Scotland, where prevailing weather conditions are favourable for field wilting, or where the enthusiasm of a successful operator has proved infectious.

A good deal of uninformed controversy has developed in the past between the protagonists of clamp silage on the one hand and tower silage on the other. We do not subscribe to the view that there is a conflict of interest between the two methods. Indeed the more one examines the process and practice of making tower silage, the more evident are the similarities with the standards and the recommendations already made for clamp silage.

Filling a tower requires a precise specification for the crop material, particularly with regard to its stage of growth, dry-matter content and chopped length. The regular daily addition of each layer of silage into the tower is also most important in controlling the fermentation that develops. A rigid application of the same discipline would probably be the most effective way of improving on the happy-go-lucky method used by many farmers who make clamp and bunker silage.

If a tower is filled with material of a dry-matter content below about 30% the outward pressure exerted on the bottom of the walls of the tower can cause a serious problem of effluent flow, and may even damage the structure of the silo. In addition, if the chop length is longer than 2 in, or 1 in in the case of some grasses, the unloading equipment may not be able to handle the silage when feeding time arrives and it will have to be dug out by hand.

POINTS TO CONSIDER

These considerations underline both the strength and the vulnerability of the tower silage concept. The cost of the tower and its ancillary harvesting, filling and unloading equipment must be financially justified by the reduction in nutrient losses, the improved animal performance resulting from enhanced storage efficiency of a high-value product, and by the greater economy of labour in a mechanised feeding system.

The need to fill the tower with material of a high production potential cannot be too strongly emphasised, and we have already shown how closely this depends on the stage of maturity (date) at which the crop is cut. Placing this consideration alongside the need for a high dry-matter content presents the basic difficulty in the production of tower silage. It is this need to combine a high inherent value in the harvested crop, with a predetermined level of dry-matter, which can make the introduction of the unpredictable weather element such a decisive factor. In the practical situation the harvesting of a particular crop will frequently demand 10 fine days, within a prescribed 14-day period, if the various related objectives are to be obtained.

This problem can be acute with the high-production, early-season grasses, which are very sappy and can be difficult to dry to the desired level in the cool and humid drying conditions common in late spring. For this reason some storage towers in the UK are now being filled in July with whole-crop cereals and in September and October with maize. With these crops the fall in digestibility of the standing crop as it matures is much less rapid than with the grasses (fig. 5), and the dry-matter content of the forage, even when direct-cut to the tower, may be high enough for satisfactory results.

If the management problem of field wilting can be overcome, there is no doubt that storage in towers is a most efficient process. But the advantages have perhaps been exaggerated by comparison of *in-silo* losses between different types of silo, whereas it is the *system* loss, between the growing crop and the animal, that really matters. Thus to the in-silo loss must be added any losses during the field processes of wilting and harvesting. Few valid comparisons have in fact been made (if two different silos are being filled with alternate loads from a harvesting team, it is likely that each silo will only be filled at half the rate it should be). Figures quoted in 1968 indicated that the 'system' loss in a tower-silo system should be considerably lower than in a well-managed bunker system; but in direct comparisons made at High Mowthorpe EHF in 1965 and 1966, the overall loss when crop of 26% dry-matter was ensiled by a system similar to the Dorset wedge (p. 109) was almost identical with that when crop of 38% dry-matter was stored in a tower silo— and this was before the recent introduction of chemical additives, which are likely to reduce losses in a bunker system.

Another advantage that has been shown for tower silage is that, because it is wilted, animals will eat more of it than of normal unwilted silage (p. 42). But with better control of fermentation by the use of additives, leading to higher intake with low dry-matter silage, this advantage may not now be as important as was found in earlier studies.

MAIN CASE FOR THE TOWER

Thus in our view the main case for the tower silo (photo 31) must be that it allows a more fully-mechanised system of feeding silage to stock than other methods yet developed. This must be exploited by siting the tower so that the unloading and handling equipment (photo 32) can convey the silage directly to the animal. In selecting the size of tower to be installed professional advice should be taken on the diameter and height to suit the particular livestock enterprise. A minimum 6 ft daily depth of fill, with preferably 10 ft, and the unloading of at least a 3 in depth of silage per day for feeding, will determine the cross-sectional area of the silo, and hence its diameter. Failure to make these calculations correctly will give rise to heating and fermentation of the exposed surface of the silage both at filling and at unloading time. This problem will be the more serious since high dry-matter material heats more readily than the wetter material generally ensiled in a bunker silo.

The special interests of tower silo operators are the concern of the National Forage Tower Council. Excellent advisory facilities are provided by the Council, as well as by tower silo manufacturers. Further information on the design and operation of tower installations is included in the Farm Electrification Handbook on Automatic Feeding and in MAFF Short Term Leaflet No. 126. The information available from these specialised sources is not repeated here; but applied by the careful operator it should allow the tower silo system to be used effectively on the medium to large-sized farm in areas where the climate offers reasonable conditions for field wilting.

Chapter 8

GRASS-DRYING

GRASS-DRYING is not a new process. It was introduced into the UK in the 1930s and by 1950 over a thousand driers were in use. But this was a time of shortage of feedstuffs, and when feed rationing ended in 1952 many of these driers ceased operating, because their production costs were too high and the value of the dried product was too low. High costs were mainly the result of small scale and low efficiency of operation (average drier output was only 200 tons, and many produced less than 80 tons); the product was of low value because much of the grass that was dried was mature and indigestible.

However, a few driers remained in operation; these were mainly the larger, more automated, units, linked with skilled field management to provide crops of high quality for drying over a long season. Most of the dried product was used by compounders as a source of vitamins and pigments in poultry and pig rations, and it was not until about 1965 that there was renewed interest in the possible role of dried grass for ruminant feeding.

The principles of grass-drying and the value of dried grass in ruminant rations have already been discussed.

Hay and silage production are taken for granted as acceptable methods of forage conservation on a farm scale. In contrast, the high capital cost of grass-drying equipment, and the need for a very high standard of management of cropping, and of harvesting and processing the crop over a long drying season, have led to high-temperature drying gradually becoming a factory rather than a farm operation. But the same care must be given to choice of equipment and management planning to make the enterprise profitable whether the drying enterprise involves a capital investment of over £100,000, and an annual output of upwards of 5,000 tons, or is planned to produce only a few hundred tons annually, with an investment of less than £10,000 in a dual-purpose grass and grain drier.

A major objective must be the provision of a succession of crops to keep the equipment working over a long drying season, since after the cost of growing the crop, depreciation and interest charges (amortisation) are often the biggest single cost in producing a ton of dried grass. Of equal importance, before deciding to install a drier the farmer must have a very clear idea of how he is going to dispose of the dried product, whether by sale to the compounder or directly to the ruminant market, or by consumption on the farm.

Provided that these questions of crop production and of product disposal can be satisfactorily answered, the overall operational efficiency and profitability then depend on two main factors.

The first is the selection of the correct size and type of drier, and of processing, conveying and storage equipment to dry and package the product in a form best suited for the planned method of use.

The second is the provision of a well-matched field harvesting system, which can cut and pick up herbage from April to November without breakdowns.

Equipment

Driers

The drier is the most expensive single item of equipment, and success will depend very much on its operational efficiency. It is likely to cost anything from £5,000 up to £40,000, and the total investment in the whole installation, including buildings, field machinery, processing equipment and storage will be 2 to $3\frac{1}{2}$ times this cost.

The choice at present lies between a rotary-drum pneumatic high-temperature drier, with inlet air temperature varying from 600° to 1100°C, or a low-temperature conveyor drier, in which inlet air will be controlled within the range 100°C to 150°C. Grass temperature of course remains below 100°C as long as water is being evaporated from it.

High-temperature driers

The most likely choice, where the main requirement is for green crop conservation rather than for grain or seeds drying, will be the rotary drum drier (photo 33). A big advantage of using high tempera-ture is that the drier is much smaller than a low-temperature drier of similar evaporative capacity; thus the design lends itself to the production of very large machines. The drum of a drier rated at 1 ton output per hour is about 25 ft long and 8 ft diameter, compared with a length of up to 120 ft for a 1 ton per hour low-temperature conveyor drier.

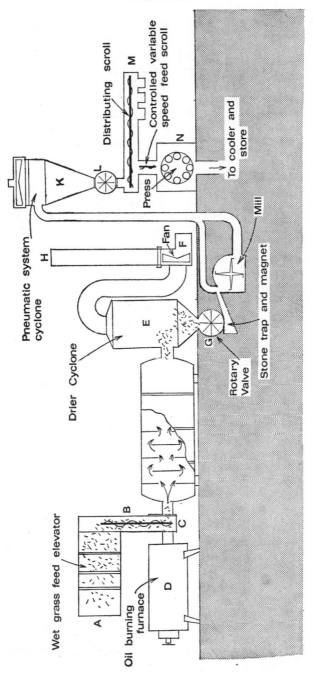

Fig. 13. Layout of a single-pass high temperature grass drier with hammer-mill and pelleting press.

Recent development has mainly been of high-temperature machines and few medium or low-temperature types are now available. A typical rotary drum drier is shown in fig. 13.

The crop, which must be chopped, is carried through the drum in the hot airstream; the weight of crop dry-matter in a 1 ton/hr drum may be as little as 50 lb at any one time. As the crop passes from the inlet (B) to the exhaust end (E) it is separated into light (dry) and heavy (wet) fractions. Pieces of leaf dry very rapidly, perhaps in as little as 20 seconds, and then pass out of the drum into the collecting cyclone (E); the wet and heavy stems fall through the airstream and are carried around with the rotation of the drum as they move slowly forward until they are dry enough to pass to the cyclone. Residence time, i.e. the time taken to pass through the drum, is likely to vary from $\frac{1}{2}$ to 2 minutes for much of the crop.

Rate of feed of crop into the drier is regulated by a feeder conveyor (A) whose speed is controlled by the exhaust temperature, usually measured in the main cyclone (E). In this way grass of high moisture content is fed in slower than a more dry crop. The dried grass is separated from the airstream in the cyclone and passes out through the rotary valve (G) whilst the moist air is exhausted to atmosphere through a fan (F) and chimney (H). Stones and metal are usually removed by a stone trap and magnet before the crop is fed into the processing equipment.

There are two basically-different designs of rotary drum. The triple-pass unit consists of three concentric tubes; the crop passes very quickly, assisted by small baffles, through a wide-diameter centre section, and then travels back towards the furnace end in the second section before returning to the discharge end inside the outer section. Short and even chopping of the grass is essential with this type of drum to avoid risk of blockage and fire.

The second type is the single-pass drum (fig. 13), in which various arrangements of baffles impede the flow of crop as it passes from the furnace to the discharge end; wet fragments of crop are moved progressively forwards with a tumbling motion which delays their removal until they are dry enough to be drawn out by the airflow from the fan. Since there are no sharp corners to be turned, longer grass, up to 8-10 in, can be dried. But there may be some loss of efficiency, and a risk that pieces of dry leaf attached to heavy wet stems will be burned away; thus short chopping is still preferred.

MOBILE UNITS

Within each of these types there are of course detailed design differences which influence the temperature distribution and the size of motor needed to power the fan. Although the power requirement can vary with different types of drier of similar capacity, e.g. from 120 to 240 hp with 3 ton/hr driers, the total power requirement in a drying installation tends to be evened out by the numerous motors needed for other equipment. Thus a drier producing 3 tons of dried grass/hr may have a total power requirement of nearly 700 hp. Most high-temperature driers are fixed installations, but a number of mobile units with evaporative capacities of up to 5,000 lb/hr (15 cwt dried grass) have been developed; their main advantage in practice has been not so much their mobility as that they can be operated in the open (photo 34)—although a covered site is often preferred.

Although it is necessary in planning the cropping programme to know the amount of dry crop which can be produced under standard conditions, the drier is in fact rated for the amount of water it will evaporate. Single units range in evaporative capacity from 3,000 lb, up to 50,000 lb/water per hour. The effect of different drying loads, discussed in Chapter 2, will be considered later, but the respective rated outputs of these extreme sizes of drier would be $7\frac{1}{2}$ cwt and $6\frac{1}{4}$ dry tons/hr. As a guide to likely requirements an evaporative capacity of up to 4,000 lb/hr is needed per 1,000 tons of seasonal output, based on the performance of existing driers operating over a wide range of crop and weather conditions. About one gallon of fuel oil will be burnt for each 100-120 lb of water evaporated, giving a specific fuel consumption of 1,350 to 1,600 btu/lb of water evaporated.

Although the rate of oil consumption per hour is a feature of the particular size of drier, and is fixed within narrow limits, the amount of oil used to produce one ton of dried grass depends almost entirely on the moisture content of the wet grass. A satisfactory average figure will be 60-70 gal/ton, but a variation from 45 to at least 120 gal/ton can be expected if crop moisture content varies from about 70% to 90% over the season. Provision must be made for the consumption of about 200 gal of oil for each hour of operation with a drier producing three tons of dried grass an hour (24,000 lb/hr evaporative capacity), under favourable conditions.

Medium-temperature driers

The medium-temperature continuous-flow drier has an inlet temperature of about 150°C and carries grass on a conveyor through which the heated air is blown. Residence time will vary from 20 minutes to nearly an hour depending on the drier design and the moisture content of the crop. Drying can be uneven if the crop has been chopped very short or is heavily lacerated; wet patches of grass can create problems in the processing equipment and in store, and will also reduce efficiency. This problem can be partly solved by using a long chop, whilst some driers have an agitator which stirs up and loosens the partly-dry grass as it passes a point along the conveyor. To improve efficiency of operation, air which has passed through the driest grass at the outlet end of the conveyor, and so is still hot and quite dry, is re-circulated by a duct to the intake of the main fan.

Evaporative capacity of this type of drier is generally less than 8,000 lb water/hr (1 ton/hr dried grass); as 90 to 100 lb water will be evaporated per gallon of fuel, maximum oil requirement will be about 80 gal/hr. The main advantage is that these machines can dry grain and grass seeds etc, without much modification, and they therefore have a place on mixed arable and livestock farms. They are also suitable for grass which has been wilted in the field, as the fire risk when drying this rather non-uniform material is less than in a high-temperature rotary drum drier.

Heavy grades of fuel oil, from 200 up to 3,500 sec, are generally preferred because of their lower cost per unit of heat—an important factor when oil prices have risen by 50% in less than three years. Oil with a viscosity rating above 950 sec must be pre-heated before it is pumped to the burner, and the heaviest grades must be delivered warm into an insulated storage tank fitted with an immersion heater.

Processing equipment

Most of the 90,000 or so tons of dried grass produced in 1971 were included at a low level in pig and poultry rations; for this it was hammer-milled and stored either as meal in sacks, or as pellets. The dried grass from the rotary valve (G) passes through the hammer mills (J) to the pelleting press (N), generally a ring-die press of the type used in feed compounding. Pellets are of $\frac{3}{8}$-$\frac{1}{2}$ in diameter, and molasses or steam can be applied to improve pellet stability. Methods of conveying and cooling the pellets are well established, and the product is suited to bulk storage and mechanical handling. For this market the dried grass is sold mainly for its content of protein, carotene and xanthophyll (the pigment that colours egg-yolks); although there is some decline in 'quality' during storage, few operators have installed the expensive gas-tight silos needed to prevent this. Slight moulding in bulk storage is restricted to the upper layers of the heap, and this can be reduced by covering the surface with plastic, and by eliminating through-draughts of moist air. Many driers, with outputs of over 2,000 tons annually, are likely to continue with this method of storage and processing.

However, consumption of dried grass by pigs and poultry is

limited. Any major increase in the amount of forage conserved by high-temperature drying must be accompanied by a similar increase in the amount of the dried product fed to ruminant animals. It may then be possible to take a different approach to processing and packaging; in particular the use of expensive hammer-mills, with their very high power consumption, may not be essential.

Some form of packaging will almost certainly be needed. The storage and feeding of loose dried grass presents many of the problems that have been discussed with barn-dried hay (p. 95) and will be adopted only if a satisfactory small-farm drier is developed. On-farm baling of long dried grass increases its density for storage, but does little to improve its handling characteristics. However, in most cases the dried product will be compressed into smaller high-density packages before it is stored or fed.

Packaging

Three basic types of packaging have been developed:

(a) 'Wafers', in which the chopped dried grass is compressed by a piston-type press, which retains a proportion of quite long pieces of dried grass in the packages.

(b) 'Pellets', produced by compressing *milled* dried grass in a rotary-die press.

(c) 'Cobs', produced by compressing *chopped* dried grass in a rotary-die press: the particles size in cobs is intermediate between those in wafers and pellets.

Wafers

Wafering presses were developed to produce packages of dried grass containing the 'long' pieces of fibre considered to be needed by the dairy cow (p. 47). Most of those in use are modifications of briquetting presses for sawdust and coal. A reciprocating piston presses the dried grass into a long cylindrical die (photo 35), producing flat 'wafers' with a preferred diameter between 2 and $2\frac{1}{2}$ in, a unit density of 35-50 lb/cu ft, and a bulk storage density of 25-30 lb/cu ft. A proportion of the particles in these wafers are longer than $\frac{1}{4}$ in, and the modulus of fineness (a measure of the distribution of size of particles) is generally between 3·5 and 4·5.

The output from a single 2 in die machine is about $\frac{1}{2}$ ton/hr, with power requirements of 20-25 hp/ton hour. To increase output, machines with two, three, or four dies have been made, and wafer diameter has also been increased up to 4 in. Double-piston presses have proved satisfactory, but feeding dried grass to more than two dies has been difficult. Wafer stability depends on crop type and quality, but all wafers tend to break down to small particles (fines) during storage and feeding, and this can cause problems with dust. Larger diameter wafers tend to have the poorest handling characteristics, and wafers over 2 in diameter are seldom made.

Although wafers are not well adapted to mechanical conveying, they can be stored in a bunker silo under a Dutch barn, and handled from this by a front-end loader and self-unloading wagon. Some wastage can occur during feeding by animals dropping partly-eaten wafers on the floor, and there is also a slight possibility of animals choking with wafers more than 2 in long. The latter problem can be avoided by reducing the length, by halving or quartering during production, and by breaking down the wafers by mechanical treatment before feeding.

However, feeding experience in the UK has not confirmed the earlier Danish evidence that wafers fed alone will give satisfactory butterfat levels in milk. In general hay, silage or straw must also be fed and it then seems sensible to feed sufficient of one of *these* forms of conserved forage to keep up butterfat level. In that case, large particle size in the dried grass is not needed, and there is in practice a real advantage in ensuring fairly small particles (p. 142). Since most dried grass will be fed in this way, as a supplement to hay or silage, some more satisfactory method of package than the wafer is needed.

Pellets

As has been noted, this is the commonest form of package at present. The storage and handling of pellets is very satisfactory, and the small particle size is ideal when the dried grass is fed in mixed rations. But the installation and operating costs of both a hammer-mill and a pelleting press are very high.

Cobs

Thus there is now much interest in cobs, produced directly from chopped dried grass without hammer-milling. Initially the aim was to produce a cob containing a high proportion of large particles; but, as with wafers, large particle size is no longer considered generally necessary or desirable, and machine design is likely to move in the direction of ensuring a high degree of grinding as the cobs are produced.

Cobs can be produced in several types of rotary press, including orbiting roller and rotating-die machines (photo 36), and flat-bed presses in which rollers force the dried grass through holes in a fixed horizontal die. These machines are all based on those used in feed compounding, with modified mechanisms to ensure an adequate feed-rate with the bulky chopped dried grass. Diameter of the cobs

POLYTHENE
SHEETING

25. Method of fixing a 6 ft. strip of 300 gauge polythene sheeting to the silo side-wall, to ensure adequate sealing of the shoulders of the clamp.

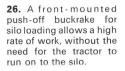

26. A front-mounted push-off buckrake for silo loading allows a high rate of work, without the need for the tractor to run on to the silo.

27. Addition of a grapple fork to a buck-rake loader increases the weight carried to nearly a ton and can improve loading rate to double that with a buckrake.

28. As each section of the silo is filled, the side-sheeting (photo 25) is pulled across the shoulder and the top sheet is stuck to it, using mastic.

29. A 500-gauge polythene sheet being placed in position across the end of an outdoor wedge-clamp silo, within 2 days of loading.

30. A good covering of farmyard manure ensures that the plastic sheet is kept in close contact with the silage.

31. A glass-lined steel tower silo being loaded by blower. A high rate of loading is ensured by the provision of adequate power to maintain speed of rotation of the fan and thrower unit, and by rapid emptying of trailers by the use of elevated platform and flared trailer sides.

32. On the farm of Mr G. Munro, in Scotland, high dry-matter silage, unloaded from the bottom of a tower, is blended with barley and fed directly once a day to 200 beef cattle using a belt feeder. A double feed is given on Saturday, so that there is no need to feed on Sunday.

(and pellets) can be from $\frac{1}{4}$ to $1\frac{1}{2}$ in, with feeding experience indicating a preferred diameter of $\frac{1}{2}$ to $\frac{3}{4}$ in, with a unit density of up to 70 lb/cu ft, and a bulk density of 50 lb/cu ft. Mechanical conveying is most effective if the length of the cobs is about twice the diameter.

Some grinding is an inherent feature of the 'cobbing' process, resulting from the metal-to-metal contact and skidding and slipping between die and rollers. This grinding, coupled with the friction in the dies, accounts for the high power requirement, up to 50 hp per ton/hr, with most machines. The amount of grinding increases with the speed of rotation, and is greatest with immature crops and with crops dried to low residual moisture levels. The density and 'stability' of cobs increases with the degree of grinding: coupled with the nutritional advantage of small particle size this indicates the likely trend in machine design. However, cobs are always likely to contain coarser particles than pellets, with a modulus of fineness in the range 1·5 to 3·0, compared with 0·5 to 1·0 in pellets.

Whichever type of press is installed, its capacity must be sufficient to deal with the maximum likely rate of output from the drier; output in mid-summer can often be twice that with the very wet grass harvested early and late in the season. Thus a machine with too low capacity may restrict output when drying conditions are favourable; on the other hand, too large a machine, besides being very expensive, may not work efficiently because it is difficult to feed the dried grass evenly into the die.

All types of package must be thoroughly cooled to prevent 'sweating' and moulding in store; this can be done in a ventilated trailer, or in a horizontal cooler (see fig. 13). This must be as near the press as possible, as cobs and pellets are very fragile until cooled, and are easily damaged by equipment that handles them roughly.

Harvesting grass for drying

The problems of mowing grass for drying are much the same as for haymaking and silage, and some of the equipment discussed in Chapter 5 will be used. Difficulties arise, however, because harvesting must continue over many months under a wide range of weather conditions, and at times when both crop and soil are very wet. Machinery must therefore be extremely robust.

Before selecting from the available systems of harvesting and transport it is essential to appreciate the full size of the handling problem. Table 9 (Appendices) shows the amount of grass that must be carted from field to drier for four different drier sizes and at two levels of crop moisture content. Grass directly cut and loaded will contain over 80% of moisture, except during very dry weather.

E

Field wilting is needed to reduce moisture content below 75% and this can substantially complicate the organisation and management of the harvesting operation.

For the smaller sizes of drier a well-organised harvesting system, as operated on many farms for silage-making, and using a trailed precision-chop forage harvester, will be satisfactory. Delivering 10 to 15 tons of fresh-cut grass per hour, harvesting will normally continue for 10 to 12 hours per day.

However, the problem is more complicated than with silage. Grass for drying is often cut when dry-matter yield is below $1\frac{1}{2}$ tons/acre, compared with two tons or more for silage; so the harvester must travel further to collect each ton. Transport distance from field to drier is generally longer and more variable. Further, the amount of wet grass needed per hour varies with its moisture content. Thus balancing the trailer force to the output of the harvester demands considerable skill. This is reflected in the percentage of the available time (ignoring breakdowns) that the harvester can be cutting; with silage-making this is often over 80%, whereas with grass-drying it is seldom over 70%, and may be as low as 50% when heavy yields of wet crop are being cut. So a system which is well on top of the job in April and May can be fully stretched in July, when light, fairly dry, crops are being harvested.

Forage harvesters, working long hours daily for several months, are subject to considerable mechanical breakdowns; these are worse when grass, compared with lucerne, is being cut. To avoid the drying operation being held up by breakdowns, harvesting and transport systems for grass-drying tend to be 'over-capitalised' compared with those for silage-making.

Most handling systems use the precision-chop harvester although a few driers will deal with the long grass from flail harvesters. Short-chopped grass packs at high density in the trailer (Table 8), important when haulage-distance is greater than a mile; it also flows freely through the feed mechanism to the drier.

For large driers, needing upwards of 150 tons of fresh grass daily, the self-propelled forage harvester, with a cutting width of 8 to 12 ft, and with a 100 to 200 hp engine, is generally used. These machines can mow, chop and load at 15 to 25 tons/hr, depending on crop type and yield, and can harvest between two and four acres an hour. Because of fairly frequent breakdowns two harvesting machines are needed to ensure the supply of crop to the drier; where only one self-propelled harvester is purchased, at least a trailed harvester will be needed as a standby. Repair costs are high, possibly 10% of the capital cost per annum, compared with 3% with harvesters for silage.

Transport systems

Two systems are used for transporting the cut grass. In the first the grass is loaded directly into the trailer in which it is taken to the drier. However, the trailer must be taken on to the field, and this can limit its size, particularly in wet weather. Thus in an alternative system the cut grass is loaded into a high-tipping trailer which holds three or four tons of crop. Two of these trailer-loads are then tipped into a road-transport trailer (photo 37) parked at the side of the field, for high-speed haulage to the drier.

Detailed studies of grass-handling systems show the difficulty of achieving a balance between the trailer capacity and the rate of field harvesting, particularly where there are frequent changes in crop yield and in the distance from field to drier. Some inefficiency in the field harvesting system may well have to be accepted so that the crop supply to the drier is assured—and also to allow adequate rest breaks for staff who must work up to 12 hours a day, six or seven days a week, for several months. Careful planning of this vital phase in crop drying is essential. The correct choice, and method of operation, of equipment will be helped by recent computer studies, in which field layout and cropping are considered in relation to drying load and drier throughput.

An alternative method of harvesting uses a flail harvester and trailer combination in the field (photo 38), and a stationary precision chopping unit at the drier. This system has advantages where transport distance is short, but is less suited for long distances because the trailer holds less dry matter as flail-cut than as chopped grass.

Several methods are used for handling grass into the feeding mechanism at the drier. In some cases the trailer can be tipped directly into the feeder, but it is more common to tip on to a concrete apron. This can serve as a buffer storage area, sufficient grass often being tipped during the day to keep the drier operating overnight. The grass is then fed into the drier with a tractor and front-end loader; loading is a rugged operation, and an industrial tractor, or at least a tractor with strengthened transmission, is needed.

DRIER OPERATION AND MANAGEMENT

Much information is available from manufacturers, and from other sources, on the different types of drying and processing equipment available, and on the harvesting and transport machines needed to keep it working over long periods. However, if the different phases of the operation are not matched in a well-balanced system output and profitability can fall well below expected levels. Certainly during the first season of operation the multiple problems of growing grass when it is needed, of harvesting and transporting it without breaks, and of operating the drier (nominally automatic, but in practice responding readily to operator competence), sometimes seem insuperable! A brief account of some of the factors limiting throughput may help in this respect.

The skilled operator aims to produce a certain number of tons of dried product from a given acreage over the season. First he must plan his cropping programme to ensure a steady supply to the drier, using information such as that included in Chapter 4; he must then harvest and transport the crop at the rate the drier can deal with it. He has some control over the rate of throughput of the drier by attention to details such as the relationship between crop type, feed rate, exhaust temperature, and final moisture content of the dried product. But the main factor controlling output is the moisture content of the crop, which varies considerably at different times of year (Table 10).

Just as there is variation over the season, so will there be variations between days, and between parts of the same day, all of which must be considered when deciding the rate at which crop is to be harvested —a heavy fall of rain on a dry day can reduce drier output by as much as a third. Most driers are rated for grass at 80% mc; in practice mean seasonal moisture content can be over 82%, and a drier rated at 1 ton/hr will then produce on average only 0·88 tons, a reduction of over 12%. Under these conditions a 1-ton per hour drier would produce only 2,600 tons in 3,000 drying hours, compared with the expected 3,000 tons.

WILTING BEFORE DRYING

The theoretical advantage of wilting all grass to 75% mc (equivalent to 'July' grass) is thus clearly seen; it would increase the output of the one ton drier to over 1·3 tons/hr, an increase of over 30%. However, this is seldom achieved, partly because of the difficulty of wilting under many field conditions, but also because unless grass is wilted very uniformly, it easily scorches and inflames in the drier; when this occurs the inlet temperature must be reduced, and some of the gain from wilting is lost. This is best avoided by taking care to get uniform wilting, and by very fine chopping, so that each single piece of grass does not contain both 'wet' and 'dry' fractions. The advantage of a mechanical means of expressing water has already been noted (p. 77); no commercial equipment is yet available, but its effect on drying costs could be so great that its development merits serious study. Thus fuel consumption at 75% mc would be 51 gall/ton, compared with 79 gall at 82% mc, and capital and labour costs would also be reduced.

However, it is essential to note that, if wilting or dewatering are practised, then the packaging and cooling equipment must be big enough to deal with the increased drier output.

The mean monthly figures given in Table 10 of course include days when crop moisture content is above 90%. Although it is considered good practice to spread capital and labour costs by drying for as many hours as possible during the season, the wisdom of drying when output is reduced to 45% of the rated level, and oil consumption is up to 160 gall/ton, must be questioned. Under these conditions the drier should probably be shut down.

The final moisture content of the dried product, which can be controlled by the skilled operator, can also affect output. Drying to 12% mc instead of 8% can increase output by up to 6% and reduce oil consumption by over 3 gall/ton. Short chopping also helps the crop to flow uniformly through the drier, and an increase in chop length from 2 in to 6 in can use up to four extra gallons of fuel per ton of product.

SOME COST ASPECTS IN GRASS DRYING

Grass drying is a conservation system of high capital and high running costs. To be profitable it must be operated with great efficiency and the product must be of high nutritive value so as to command a good market price. Although direct costs may vary widely under different conditions, an indication of likely levels can be given (Table 11).

Ways of reducing *direct costs* on oil have already been noted. Equally important is the effect on costs of the scale of operation, as overhead charges increase substantially with any fall in drier output. The figures in Table 12 have been produced by a Lincolnshire drier operator, based on his own experience with drying grass and lucerne.

Allowing £10.20 per ton for the crop, the total production cost with the largest drier could be £20 per ton if it worked at full capacity, and nearly £23 per ton with the smallest drier, working at 80% capacity. Other reports* have shown somewhat higher costs, with dried lucerne (4 tons/acre) at £26/ton, and dried grass, despite a yield of 5 tons/acre, costing £27/ton because of the £3/ton fertiliser cost. In these cases the write-off period for the drier was eight years, and for the forage harvesters only four years.

As has been noted, such costs can only be approximate, and other aspects such as the financial policy and tax position of the business

* Dried crops for ruminant feeding. Survey: Dri-Crop Developments Ltd.

can also affect production costs, and so the price the dried grass must realise to be profitable. But it is clear that quite small improvements in crop yields, in harvesting and drying practice, and in the way in which the product is fed, can have a significant effect on overall profitability.

What does emerge however is that grass-drying is an enterprise for the specialist, able to operate on a large scale and to employ the skilled labour needed to get the best out of complicated harvesting and drying machinery. It is not the conservation system for the 'average' livestock farmer—unless he is prepared to link with other like-minded men, to set up a drying co-operative big enough to ensure economy of scale. This has become common practice on the Continent, particularly in France and Denmark; but most of the co-operatives are planned to dry crops for small farmers (one unit in Denmark harvests grass from more than 400 farms). By contrast, less than ten farmers are involved in each of the co-operatives so far set up in this country. These have mainly been arable farmers, wanting to dry the forage break crops they consider essential to maintain soil structure and fertility, and in several cases selling the whole of their production off the farm.

Some of these producers also feed part of the product to their own stock. But for most livestock farmers barn hay-drying and ensilage will remain the ideal methods for conserving surplus grass during the grazing season; for these, cutting can be concentrated during only a few weeks, in contrast to the 180 days or so needed for economical operation of a grass drier. But the livestock farmer will still need to purchase other feeds to supplement the hay and silage he makes—and it is here that the main future role of dried grass seems to lie. As discussed in the following chapter, dried grass cobs and pellets, probably mixed or compounded with mineralised barley, can be a most effective supplement to other forms of conserved forage. Thus dried grass and barley, produced as cash crops on the mainly-arable farm, will be exported to neighbouring livestock areas (movement of dried grass from one side of the country to the other will be discouraged by high transport costs).

THE GRADING OF DRIED GRASS

Arrangements for grading and marketing dried grass will then be needed. At present dried grass is sold under a guarantee of protein content, and in addition some purchasers stipulate minimum carotene levels. For ruminants, however, the digestibility (D-value) of dried grass is also important, and this has led to suggestions that multiple grade levels, based on both protein content and digestibility, will be needed. But this would increase the complications of storage, distribution, and feeding (each of perhaps 12 or 14 different grades would have a different feeding recommendation), and any system adopted must be based on the least possible number of grades. Experience at a large unit in Denmark suggests that one solution may be the bulk storage of a mixture of the different lots of dried grass produced during the season. After cooling, the dried pellets are evenly distributed by a shuttle conveyor on to the already-dried material in store; at the end of the season this store contains some 10,000 tons of pellets (photo 39), and each lorry-load taken from the stack is of virtually the same analysis.

There may be advantage in separately storing the highest-quality lots of dried grass produced in early spring, and possibly the very low-quality grades likely to be made at times in mid-summer. The remaining lots, of differing protein and digestibility levels, and cut from a range of forage species, would be mixed to produce a 'standard' dried grass feed, and the effects of small variations between lots would be reduced by mixing the dried grass with barley before feeding.

Some such system of standardisation seems essential if the producer and the feeder are not to be confused by a multiplicity of grades. Dried grass would then be sold on the basis of a minimum protein content (say 16%) and a minimum digestibility level (say 63% D-value, but possibly lower where most of the crop dried is lucerne); the price charged and the method of feeding adopted would be based on this quality.

'Standard' dried grass could then play an increasing role as a reliable feed supplement. Both arable and livestock farmers must now examine the possible impact of this new form of conserved forage on their farm economies.

Chapter 9

FEEDING CONSERVED FORAGES

MANY TECHNICAL books and leaflets give recommendations on the feeding of hay and silage in the rations of different classes of livestock. In most cases these are concerned with forages conserved by fairly traditional methods; field-made hay (at over nine million tons a year, by far the main winter forage fed in the UK); and trench, clamp or bunker silage. Conservation losses by these methods have generally been accepted as running above 30%, and the hay or silage has often contributed only maintenance feeding to the total ration.

We believe that these low-to-medium quality feeds, made with high losses, will have less role to play in future feeding systems and that, if advice *is* needed on how to make them and feed them, this is already easily available. Our concern here will be with the feeding of the better quality conserved feeds described in earlier chapters—hay and silage made from forages cut at a not too mature stage, conserved by practical methods which minimise losses, to give feeds of good intake and digestibility characteristics. Only this class of feed can make up a high enough proportion of the ration of productive animals to reduce the level of supplementary feeding with cereals and other concentrates. As we have noted earlier, this will become more important if the price of these supplementary feeds rises more rapidly than the farmer's receipts for milk and meat—as seems likely to happen within the EEC.

FACTORS TO BE CONSIDERED

But several factors, varying between different farms, must be considered. Thus in aiming for greater use of conserved forages, milk yields or daily weight gains must not suffer, for this could cancel out any advantage from cheaper food costs. Further, the need to put aside more acres for conservation must be considered in the context of the overall stocking rate on the farm. In the 1960s it was often more profitable to reduce the amount of forage conserved, so that more cows could be grazed in summer, and to purchase much of the winter feed. But on many farms stock numbers are now up to the limits determined by housing and labour.

Grazing will remain the main method of summer feeding, but attention can now more seriously be given to feeding more conserved forages in winter so as to reduce purchases of feedingstuffs. The first feature that emerges is that, if conservation losses can be reduced, then more hay or silage will be available for feeding to each animal kept. Secondly, once more effective methods are adopted there is then a real incentive to grow bigger crops for cutting. There is little doubt, for instance, that levels of nitrogen fertiliser applied to hay-fields are often low (or non-existent) because it is so much easier, with traditional methods, to make hay from a light crop. And, particularly with the use of additives, it is now possible to make good silage in the autumn from surplus grass which previously has often been wasted.

The first step must be to grow higher yields of grass and other forages, both for grazing and for cutting. Decisions on the types of crop to grow will be helped by the new information on yield, digestibility and protein content, and this will also indicate the best time for cutting.

STORE CROP WITH MINIMUM WASTE

The aim must then be to store the crop with as little loss or wastage as possible by one of the methods described here—quick field hay, barn-dried hay, sealed bunker or clamp silage, tower silage and, in a few cases, as dried grass. Before winter feeding begins, an estimate is then made of the approximate amounts of the different qualities of conserved feeds available, and a feeding plan worked out to decide how these will be fed to the different classes of stock—always with the aim of ending the winter with at least a few bales of hay in hand! This plan must, of course, include the other foods available on the farm (straw, sugar-beet tops, kale and roots, and home-grown cereals), and will also consider which bought-in supplements will represent best value, and whether home-mixing or contract-mixing will cheapen the cost of any compounds needed.

The following sections are concerned with some aspects of drawing up these plans. In this, two main alternative situations are considered —that only cereals and conventional compounds are available; and that, in addition, dehydrated forages are also available. The latter will only be generally applicable if there is a considerable increase in the UK production of dried grass—and this will only come about if there is a clear demand for this product from enough livestock farmers.

FEEDING YOUNG STOCK

Calves and early-weaned lambs are generally fed mainly on concentrates, plus small amounts of leafy hay; forages, either grazed or conserved, have played little part. With the usual quality of hay or silage this has been a wise decision; the digestibility and the intake of these feeds by young ruminants are much too low to provide adequate nutrition. Better conserved feeds, especially barn-dried hay and dried grass, can contribute much more. Thus at Hurley nearly 1,000 calves (both steer and heifer) have been reared on dried grass wafers or pellets as the only solid food fed with milk substitute. Daily gains to 10 weeks have been similar to the target figures quoted by MLC, but whether this ration cheapens rearing costs will depend much on the price of milk substitute.

More generally applicable is the feeding of hay or dried grass to weaned calves or lambs. If cut at D-values above 65, both these feeds have given daily gains of $1\frac{1}{2}$ lb with calves from 10-20 weeks of age. But at this stage, as with the feeding of most livestock, it is quite unnecessary to push the feeding of conserved forage to the limit, and sensible to include some rolled barley or weaner pellets in the ration. It is also valuable to feed at least some of the forage in the long form, so as to stimulate early rumen development and a better capacity to utilise forages.

At this young age silage has been little used, but limited experience with some newer forms of silage—of higher intake potential— indicates that these may now play a useful role in the feeding of even very young animals. Of particular interest is the recent ex- perience (p. 144) with maize silage, supplemented with protein, which has given gains by young cattle well up to the target rates of gain in most beef production systems, and at a cost well below alternative feeds.

It is also accepted that young cattle will seldom make satisfactory gains when grazing even good pasture; this is almost certainly because their grazing intake is too low, and this indicates the need to feed supplements to increase total intake. Dried grass pellets have given equally good increases in gains by grazing calves as has rolled barley. We now urgently need a study of the use of a dried grass/barley combination as a pasture supplement; this could be of particular interest later in the summer for feeding to lambs which would otherwise be sold off as stores.

CONSERVED FORAGES FOR MILK PRODUCTION

As has been discussed above, even when feedstuff prices rise, the first priority for most dairy farmers will still be to keep as many productive cows as their staff and buildings can cope with. Only then will they turn their attention to growing as much as possible of the feed for the dairy herd. But with the best will in the world it is most difficult, and probably inadvisable, to aim for self-sufficiency. The 'lactation' yield of the dairy cow is very largely determined by the level of peak yield attained at 6-8 weeks, and most recent research and experience indicates the advantage of generous feeding in the weeks before and after calving, so as to ensure a high peak yield. This is most difficult to achieve on even the best hay or silage, and the use of more concentrated supplements during these weeks is essential in the feeding of most dairy cows.

At this time the main factor limiting the use of forages is the cow's inability to eat enough of them. Even when highly digestible forage is fed, intake is still likely to be inadequate, and some less bulky, more concentrated feed supplement is needed to get a nutrient intake adequate for the demands of the cow yielding over 4-5 gallons. And it is here that the problem arises of the interaction between forages and concentrates, noted in Chapter 3. Thus a $11\frac{1}{2}$ cwt Friesian cow, eating 35 lb of DM as hay or silage of 64% D-value (45% SE, Table 4—Appendices) would be getting about 16 lb SE, enough for maintenance plus three gallons. This may seem a high potential for silage; but in 1965 the GRI, in joint work with the Berkshire Institute of Agriculture, fed Friesian cows entirely on silage, with the results summarised in Table 13; individual daily yields were over three gallons despite the rather low intake of the silage.

We believe that silage (and hay) of even better quality can now be made on most dairy farms. But in commercial practice a peak yield of only three gallons would not be acceptable. Lead-feeding, as developed at the NIRD, is therefore being adopted—that is, on many farms increasing amounts of concentrates are fed in early lactation up to the point at which daily yield increases no further, say to a peak yield of seven gallons. In theory this should require about 18 lb of No. 1 dairy nuts (12 lb SE), in addition to the basal (M+3) silage. But the cow getting this quantity of dairy nuts is likely to eat some 10 lb less silage dry-matter than it would if the silage were being fed as sole feed, so that the silage eaten is now contributing less than (M+2) to the total ration. To reach seven gallons an extra 4 lb of concentrate will thus be needed, further reducing the amount of silage the cow will eat. Because of this *replacement* of both hay and silage as the level of concentrate feeding is increased, we find that even top quality conserved forages seldom contribute more than (M+1) to the ration of the high-yielding dairy cow—although their analysis would indicate a much higher feeding value.

EXPENSIVE FEED WASTED

This would not matter—for the attainment of high peak yield is the priority—were it not that this disappointing 'potential' for the conserved forage, established in the first weeks of lactation, is so often taken as a measure of its real potential. In the above example the silage is taken to be worth only (M+1); yet when the cows are yielding (M+3) later in the lactation (a level which the silage, supplemented with a few pounds of rolled barley, could readily sustain), the 'regulation' 4 lb of concentrates per gallon over (M+1) are still being fed, and expensive feed is being wasted.

Cereals and concentrate feeds tend to reduce the intake of the forages with which they are fed as supplements. Unless this feature is recognised, we shall not exploit fully the high digestibility and high intake potential, which earlier chapters have suggested we should aim for in our conserved forages. How then can we exploit this potential?

The first course must be to make allowance for this effect of our present concentrate feeds. Just as lead-feeding with concentrates is an essential procedure early in lactation, so could lead-feeding with conserved forage be later in lactation. From the D-value of the forage a pretty good estimate can be made of its yield potential, and the level of supplementary feeding could then be steadily reduced as the cow's yield approaches this level. Even when yield falls below this it will still be sensible to feed some supplement—probably mineralised barley—to get into the ration the phosphorus, magnesium and perhaps trace elements which may be lacking in the silage, as well as some readily-available energy to keep up milk SNF levels and to avoid any risk of ketosis. But even then the conserved forage will in practice be giving around (M+3), a far cry from the Maintenance accepted in many feeding systems.

MORE EFFECTIVE USE OF PROTEIN

A particular point of economy that can emerge here is in the more effective use of the protein in the forage. Conventionally hay and silage have been cut at a pretty mature stage, and hay in particular has generally contained less than 10% of protein; as a result these feeds have had to be supplemented with high-protein concentrates to get the overall 14% CP needed in the ration of the high-yielding cow. But earlier cutting and more use of legumes, coupled with better conservation, should allow higher protein levels in conserved forages—and a reduction in the protein content of the supplements fed with them. More recently this has been accepted in the case of silage, with the advice that barley is adequate for the first couple of gallons of milk because of the surplus protein above maintenance needs contained in the silage. This has not always worked out in practice, because intake factors (Chapter 3) can reduce the amount of silage (and so the amount of protein) being eaten to below the expected level.

The production of silages of higher intake potential (by the use of additives and by wilting) should allow this greater use of barley, even in the initial challenge period, so as to reduce feed costs. Exactly the same advantage can be expected from the higher levels of protein in barn-dried hay—in fact this advantage *must* be exploited if the rather higher costs of making barn-dried hay rather than field hay are to be fully recouped.

Accepting then that a minimum of 4 lb of mineralised barley will be fed daily (reducing maximum hay or silage intake by perhaps 2 lb DM) Table 14 indicates the approximate levels of milk production (3.5% fat) at which no other supplementary feed should be necessary, with forages at different levels of D value. Note that at lower levels of D value the forage is likely also to be low in protein, and that the barley will need to be replaced by a balanced compound. But these data indicate a much lower level of supplementary feeding, particularly in the second half of the lactation, than is currently the practice on many dairy farms—and therefore enhanced profitability at this stage of lactation.

DRIED GRASS IN DAIRY RATIONS

But such systems still do not make full use of the feed potential of these better conserved forages, because of the tendency for both the cereal and the compound supplements to reduce the amount of the forage that dairy cows can eat. It is here that the different reaction to the feeding of dehydrated forages, described in Chapter 3, could be of increasing interest. As was noted there, ruminants apparently can eat more hay or silage when these are supplemented with dried grass than with the same quantity of, say, rolled barley. The main reason for this appears to be that the dried grass does not make the rumen as acid as when barley is fed (so that the fibre in the basal forage is better digested), and also that it is not eaten as rapidly. Because of this, a combined silage/dried grass ration can be as productive as a silage/barley ration, even though the dried grass is of lower unit energy value than the barley. The advantage of small particle size in the dried grass (fed as cobs or pellets), which allows the cow to eat more silage than when wafers are fed, has also been noted. Table 15 reports the levels of milk production by Friesian heifers fed grass silage *ad lib*, plus the same controlled amounts of dried grass wafers or pellets: these amounts were in fact rather lower than had been planned because some heifers would not eat their full ration of wafers, so that overall levels of milk yield were not as high as expected. But the results show the higher amount of silage eaten, and the extra milk produced, when pellets were fed.

However, as has been observed earlier, there can be advantage in using limited quantities of cereals in the total ration, particularly if this can be done without seriously reducing the amount of hay or silage eaten. In any case the inclusion of barley will be necessary, for within the foreseeable future the quantities of dried grass likely to be available will only be enough to feed relatively few dairy cows as much as 12 lb per day. Thus main interest now centres on the feeding of combinations of dried grass with barley and minerals as a supplement to hay or silage for milk production, so as to make the most effective use of a limited supply of dried grass.

MINISTRY EXPERIMENTS

Here the results already available are encouraging. An extensive series of experiments carried out at the Ministry's experimental husbandry farms are summarised in Table 16. In these a 50:50 combination of dried grass (or lucerne) with barley and minerals completely replaced a conventional compound fed as a supplement to cereal silage or to a straw balancer ration. Cobs and pellets were also compared, but there was little difference between them in milk production, so that the mean results are given. Earlier work had shown that 5 lb of high D-value dried grass was needed to replace 4 lb of dairy compound, indicating that 4½ lb of the 50:50 dried grass/cereal mix would be needed to do the same. Unexpectedly, the EHF experiments showed that only 4 lb of the mix was needed to replace 4 lb of the 17% compound, except where lucerne of very low quality, 51% D-value, was used (not reported). Most important, butterfat levels, although lower than on the control diets, were satisfactory even when the milled dried forage was fed, because the basal ration of straw or silage provided an adequate intake of the 'long fibre' needed for proper rumen fermentation.

A criticism of these experiments is that the comparison between the dried grass/barley mixture and the dairy compound was only made from the 12th week of lactation and provided no evidence of the ability of dried grass to establish the high peak level of yield in the first critical part of the lactation, which largely determines the total lactation yield. Further, the quality of the basal rations was low, and they were also fed at restricted intake for 'maintenance'. Thus the possibility of the forage also contributing to the 'production' ration, and in the case of silage, intake being higher with the dried grass/barley supplement than with the compound supplement, was not investigated.

These features need more detailed study before the role of dried grass in dairy-cow feeding can be fully established. Certainly in early lactation, as was discussed earlier, it may be advisable to *restrict* the amount of conserved forage which is fed to potentially high-yielding cows, and to 'lead-feed' high energy concentrates to establish a high peak yield. But from this point onwards it should be possible progressively to replace the concentrate with the dried grass/barley combination. Not only is this likely to be cheaper than the concentrate (even at 1969 prices it was as cheap as the home-mixed compound in the EHF experiments), but it should also allow the maximum contribution from hay or silage in the last two-thirds of the lactation.

Already some dairy farmers have begun to feed dried grass and barley as a compound replacer. However, even in pelleted form this combination seems to be eaten more slowly than conventional compounds, so that it is less suitable for feeding during high-speed parlour milking. Some of the methods being adopted to overcome this problem are discussed in Chapter 10. To date these mainly involve trough feeding of rolled barley and grass or lucerne pellets.

These methods could be greatly improved by the commercial availability of combinations of dried grass with barley and minerals. This could be in compound pellets, such as those being fed in the MLC bull performance tests (p. 146), but a mixture of dried grass pellets and mineralised barley pellets may be almost as effective, and would avoid double processing of the dried grass. However, neither of these feeds will become available without a clear indication that dairy farmers want them. And, as we have already noted, this demand will come only if there is good evidence for real economic advantage in the greater use of these feeds; we believe this advantage will become more apparent as conventional supplementary feeds become more expensive.

CONSERVED FORAGES FOR BEEF PRODUCTION

Feeding for beef has seen dramatic changes during the last decade, set in train by the development of 'barley beef'. During the 1950s a few farmers had developed methods of feeding beef cattle mainly on cereals, but it was the work of Preston at the Rowett Research Institute that put the system on a practical basis that could be widely adopted—so much so in fact that by 1963 many people considered that most of the cattle in Britain would soon be fed entirely on barley. But for a number of reasons this did not happen. Higher calf and barley prices and below-optimum gains cut margins on many farms, while the 'premium' market for barley-beef seemed to be limited. The main reason, however, was the development of alternative systems of feeding. Of these the most widely adopted has been the '18-month' grass-beef system. In this the rather lower daily gains and the slower turnover of capital are more than compensated by feed costs being less than with barley, and by the higher market return from the larger carcass produced.

In fact 'barley-beef' was the best thing that could have happened for 'grass'. It pinpointed the weakness in earlier traditional grass-beef systems, and stimulated the application into practice of better techniques of both grazing and winter feeding of forages. In particular it was found, by feeding good quality hay or silage with only limited amounts of cereals, that gains in winter were much higher than in the conventional 'store-feeding' period, so that cattle could be marketed at 18 months instead of at 2-2½ years of age. Even better levels of winter production should now be possible using some of the recent improvements in conserved forages.

MAIZE SILAGE FOR CALVES

The role of silage and hay in the feeding of young cattle has already been noted. Traditionally these feeds have played little part in this rearing stage because their digestibility and intake have been so inadequate. But barn-dried hay is being fed successfully on the farms where it is made, and limited experience with feeding the newer 'high-intake' silages to calves is encouraging. Some of the most interesting results have been with maize silage. Both the intake and the protein content of this silage are too low for good gains by calves: but in recent work at Hurley three month old steer calves have made gains up to 2 lb per day when fed on maize silage supplemented only with urea and lucerne pellets. The urea, *mixed uniformly with the silage before it was fed*, made a useful contribution in cheapening the total cost of this ration up to about nine months of age; beyond this age the protein content in the particular silage fed (10·6% CP) appeared to be adequate for satisfactory gains (Table 17).

However, while maize silage is likely to be fed in the southern part of the country, winter feeding in most areas must be based on conserved grasses and legumes. Clearly the exact requirement will depend on the particular beef system adopted: the spring-born calf fed in its first winter before being finished at grass may not need to gain as rapidly as the autumn-born calf which is to be fattened

33. High-temperature triple-pass rotary drum drier. Models are available with an evaporative capacity from about 3,000 lb to more than 40,000 lb of water an hour.

34. Mobile high-temperature drier enables the drying site to be moved near to the crop to reduce transport. Output ranges from $\frac{1}{2}$ ton to about 18 cwt/hr, the limiting factors being evaporative capacity when grass has a high mc and output of the integral wafering press at low mc.

35. Piston-operated wafering press (10 cwt per hr output) installed at G.R.I. Hurley to produce wafers of 2-in diameter from chopped and ground grass, for feeding experiments.

36. Rotary-die press with 4 rollers and high speed auger-feeder designed to accept chopped unground dried grass, and to form cobs of $\frac{1}{2}$ to 1-in diameter, depending on die size. Output can be $2\frac{1}{2}$ ton/hr or more.

37. Two-stage system of harvesting and hauling grass to the drier. 3 or 4 tons are loaded into the high-lift tipping trailer, and than transferred to a road trailer, of 7 to 8 ton capacity.

38. Combination of a 52 in. in-line flail forage harvester and 9-ton 4-wheel trailer designed to pick-up wilted grass from large swaths.

Bulk storage of dried grass pellets : Shell Farm, Denmark.

40. Large bales can be fed from purpose-built feeders in field or barn, by placing into position and cutting the strings.

indoors during its second winter. The former, fed entirely on hay or silage of 60 plus D-value, can gain at a satisfactory $1\frac{1}{2}$ lb per day; the latter, which must gain at 2 lb a day or more, will need a high intake of 65 plus forage, and possibly limited supplementary feeding as well.

DRIED GRASS PELLETS WITH SILAGE

Results both from experiment and from practice have shown that high digestibility barn-dried hay will give this rate of gain. There is less experience with silage fed alone, because the information on silage intake, and on the improvement of silage intake by suitable additives, is so recent. But as with other forms of livestock production it is quite unnecessary for conserved forage, fed by itself, to be a complete feed. The aim must be to conserve forages that will make a much bigger contribution to the total ration than has been the case previously, but there may still be advantage in feeding limited levels of supplements with the hay or silage fed to beef cattle. Here again the interaction between dried grass pellets and silage may be important. Table 18 records rates of gain by cattle fed grass silage *ad lib*, together with limited supplements of dried grass pellets or rolled barley. Because the cattle were able to eat more silage when pellets rather than barley was fed, equal gains were made with a lower level of supplementary feeding with the dried grass.

Again, however, the amounts of dried grass available in the foreseeable future will not be enough to feed many cattle at this level of input, and interest must now turn to the dried grass/barley compound as a supplement to hay or silage for feeding beef cattle. Older cattle have relatively low protein requirements and the compound for these cattle could well include the low-protein but high D-value dried grass produced from Italian ryegrass or dried forage maize, particularly where the basal hay or silage has a reasonable protein content; dried grass of higher protein content will be used for inclusion in the compounds to be fed to dairy cows and young stock.

DRIED GRASS/BARLEY SUPPLEMENT

At prices ruling in 1972 there was little incentive to replace barley with the more expensive dried grass in beef-feeding, so that there is little practical experience in feeding dried grass/barley as a supplement to hay or silage for beef cattle. But where this combination has been fed it appears, just as with dairy cattle, that it has an unexpectedly high nutritive value. Thus in work at Drayton EHF beef cattle were fed on three types of ration; on a barley-beef ration; on dried grass pellets; or on different combinations of dried grass and rolled barley. Rates of gain (Table 19) were better on the barley ration than on the dried grass, reflecting the higher energy value (SE) of the barley; but gains were at least as high on the mixtures, particularly on the 50:50 dried grass/barley combination. This same combination has also given very high rates of gain with the bulls fed in the MLC performance tests. A most impressive feature of these tests has been the very satisfactory feed conversion rates (5·4 lb of grass/barley cobs + 1 lb hay, per lb liveweight gain) made by bulls weighing over 1,000 lb.

Although not strictly within the subject-matter of this book, these MLC results raise a question that cannot much longer be ignored, namely the commercial feeding of beef bulls. Because of complicated licensing regulations and an unhelpful beef subsidy system, the feeding of bulls beyond the age of 10-11 months in this country is almost unknown. Yet there is plenty of evidence that bulls can gain up to 20% faster than steers, and can produce more lean meat in each carcase—important in a situation where the supply of suitable calves is likely to be the main factor limiting the production of more beef.

As with dairy cows, more study is now needed on the feeding of dried grass/barley compounds as a supplement to hay or silage for beef cattle. A particular aspect may arise in obtaining the degree of 'finish' in cattle needed by the market, and it may prove desirable to increase the level of barley used in the last few weeks of feeding; this could have the further advantage of reducing a tendency to yellowness in the fat which may occur in cattle fed mainly on forages. But overall the aim should be to feed the maximum proportion of conserved forage, feeding the more expensive barley and dried grass/barley compound only at the levels needed to get a high rate of economic gain.

EFFECT ON SUMMER STOCKING RATES

A greater reliance on conserved feeds in winter feeding will, of course, have to be set against any effect this may have on summer stocking rates. It has generally been assumed that more hay or silage will mean that less cattle can be grazed in summer. However, several new features can be considered:

(a) As already noted, the better conservation methods now being adopted should allow heavier crops to be successfully conserved, and so could encourage the use of higher fertiliser levels to grow more grass. Further, the reduction in losses which these methods give, compared with conventional methods, will also make more winter feed available.

(b) These improved methods also make the growing of special crops for conservation more attractive—in particular forage maize in southern areas, tetraploid red clover with timothy (p. 56), and well-fertilised Italian ryegrass.

(c) A feature of the '18-month' beef systems has been an emphasis on a relatively low grazing pressure, so that herbage intake by the grazing cattle is not restricted, and daily liveweight gains are high. A proportion of the ungrazed herbage is then topped off to keep up the quality of the regrowth, and so is wasted. Recent research at Hurley suggests that the intensity at which cattle graze can be increased, with no resulting drop in daily gains, if limited amounts of supplements are provided at critical times. Rolled barley and grass pellets have been found equally effective as supplements; this should make more crop available for cutting, by allowing a smaller area to be used for grazing.

(d) It is well known that cattle which have grown slowly during a 'store feeding' period in winter can make very high rates of gain when they go out to good grazing the following spring. We now need to examine whether cattle whose summer gains have been restricted by high grazing pressure can make similar 'compensatory growth' when fed in winter on high quality conserved forages, as described in this chapter. If this does occur then it should be possible to increase stocking rates by cattle in summer, and so release extra acres for cutting for conservation.

The full implications for beef production of improved forage conservation have not yet been assessed. As already noted different strategies will almost certainly apply in different systems; spring-born cattle will have different requirements from autumn-born; breeds and crosses will differ; and the needs of particular markets must always be considered, some markets wanting older, more mature cattle than others. But the adoption of a more efficient conservation method is likely to improve and simplify the management in most beef systems.

CONSERVED FORAGES FOR SHEEP

The quality of conserved forages fed to sheep has been of relatively less importance because the sale product, the lamb, is usually fed on a ration of ewes' milk and grazed grass. However this situation could change with some aspects of intensification in sheep production. Already many sheep farmers are inwintering their ewe flocks to avoid the decrease in summer grass which occurs when outwintering is practised. Most ewes are fed on poor quality hay with small amounts of supplements, but the quality of winter feed becomes more important with the trend towards higher lambing percentages. Silage is little used, and too often the silage fed to ewes has been the partly-rotted material that cattle would not eat. A problem with silage is that it must be fed out in troughs or racks, because self-feeding is not suitable for sheep; but the performance of ewes fed on well-made medium-quality silage (60% D-value) is excellent, and more feeders should consider whether the advantages do not justify some extra labour in feeding.

Another interesting development has been the use of cobs of dried whole-crop cereals (oats or barley cut at the cheesy stage) as feed for both grazing and in-wintered ewes. Particularly in hill areas these cobs have been little more expensive than baled hay, and a number of hill farmers consider that any higher cost is fully out-weighed by the convenience and lack of waste in feeding these cobs. Further improvement could come from the inclusion of low levels of minerals and urea, but this might raise problems under present feedingstuff regulations. But a market for dried whole-crop cereals could be of real significance to the commercial grass-drier, because this crop will take up drying capacity at a difficult period in mid-summer, and this market should be promoted.

All these conserved forages are likely to need supplementing both before and after lambing; if cereals become more expensive there may be a role here for dried grass or dried grass/cereal cobs based on medium-digestibility material. These same feeds, made from forage of higher digestibility, could also be useful in lamb feeding, as a pasture supplement to finish the slower-growing lambs which would otherwise be inwintered as stores. Dried grass has also been shown to be excellent for artificially-reared lambs, which have gained up to $\frac{3}{4}$ lb daily when fed pellets of high D-value grass or legume. However, whether artificial rearing is adopted will depend on factors other than the cost of dried grass.

Most of the evidence on the feeding of conserved forages, reported here, has come from experiments at Institutes or Husbandry Farms. It is often said that the conditions in these experiments are so ideal that the results cannot be translated to commercial farm practice. But we believe they can largely be repeated by any farmer who is prepared to take his forage conservation seriously. To quote just one example, in February 1972 Howard Crapp, of Lincolnshire, was feeding his 80 winter-calving cows on medium dry-matter Italian ryegrass silage, 11 lb of dried grass pellets, 15 lb of rolled barley, 20 lb of sprout waste and 10 lb of wet brewers' grains. These feeds were put into a self-unloading trailer, which mixed them well as it delivered into feed-troughs in a partly-covered feeding area. No concentrate was fed in the parlour, and average milk yield was five gallons—a good example of experimental results being applied in practice, with the feed cost at 8p per gallon.

Chapter 10

METHODS OF FEEDING CONSERVED FORAGES

In EARLIER chapters there have been references to some aspects of feeding hay, silage and dried grass. Clearly the method of storage and feeding adopted will depend much on the size and layout of the farm, and the numbers and type of stock to be fed. But the main objective of the method adopted should be that it allows the animals to obtain a high proportion of the conserved forage in their total ration—and this means that it should effectively give *ad lib* feeding.

For this, however, the feed must be of suitable potential, in terms of both intake and digestibility, for the particular animals for which it is intended. It must be made with the lowest possible losses, both in the field and in store, fed so that intake is not restricted, and handled throughout feeding so that it is not soiled or wasted.

Hay

Much hay *is* wasted in feeding each year in the UK, often because the feeder puts such a low value on it that he considers expenditure on either labour or equipment is not justified to reduce losses. Quantities of hay (and silage) are fed out directly on to pastures in winter, and in high rainfall areas much of this is poached into the ground by the stock. Some livestock farmers have already found that cobs of low-to-medium quality dried grass, although more expensive per ton than baled hay, are more economical in the long run because they can be fed from troughs in the field without waste. But most hay is fed from racks or troughs to yarded stock. The greater part of this is baled hay, and unless care is taken much of this can be pulled out and trodden by the animals. This wastage is reduced if access to the racks is restricted by the use of welded mesh, but this can easily reduce the amount of hay the animals can eat. Probably the most effective way of feeding long hay is from troughs, using yokes to prevent the animals from scattering it around, but this of course increases costs.

Hay which has been chopped before barn-drying can very conveniently be used in a partly self-feed or an easy-feed system (photo 19), provided the barn is not more than about 15 ft wide from its centre to the manger. One man can feed up to 100 head of stock in about 10 minutes, given access to both sides of the barn. A side-delivery trailer can be used to feed out chopped hay into troughs, but foreloading into the trailer can be a dusty job, while an elevator will probably need hand-loading. With this method, however, other feed components, including silage, dried grass and barley, can readily be mixed with the hay to give a complete feed, so that better ways of handling chopped hay out of store would repay study. Thus, as noted in Chaper 6, the use of hay towers *could* bring complete mechanisation to hay feeding. Feeding experience with large bales (p. 88) is limited. It seems unlikely however that these bales will be broken up by hand before feeding, and more likely that they will be foreloaded into a V-shaped weld-mesh rack and the bands then untied so that cattle can feed from both sides (photo 40).

Silage

Most silage will be made in unwalled clamps or walled bunkers, and from these it can be self-fed, easy-fed, or mechanically fed. At present self-feeding is the most widely used, but this often limits the amount of silage the animals eat. Such restriction can arise from several causes. Where the silage is made from long herbage and is heavily compacted, stock find it difficult to pull it from the face, particularly where there is any trouble with teeth, as in cattle around 20 months old; too narrow a silage face may lead to competition and bullying between animals—thus heifers may not be able to compete with older cows; and the time available at the face may be restricted because of yard layout etc. If enough care is not taken to remove rotted patches of silage this gets mixed with the good silage and reduces palatability (this of course can happen with any method of feeding, and is another good reason for ensuring that silage does not contain rotted patches).

The Agricultural Land Service (Technical Report No. 13) has advised about $4\frac{1}{2}$ in face per dairy cow with 24-hour access, increasing to 30 in where all the cows must feed at the same time. More information, however, is needed with 'high-intake' silages, where each animal may need to spend more time at the face to exploit this intake potential.

SELF FEEDING

Self-feeding is easier when short-chopped forage is ensiled, but if the silo is filled above 6 ft in depth there is some risk of collapse of the silage face if cattle are allowed to burrow into it when feeding. This is best prevented by using an electrified wire or bar at the face to control feeding, and with high faces it is also sensible to cut back the silage above 5 ft and throw it down behind the barrier for feeding. This is particularly advised when tombstone barriers are used at the face; it may also help to reduce any limitation on intake when long silage is fed.

An incidental advantage of the Dorset wedge system is that, because the forage is not heavily consolidated during filling the silage tends to be less dense than is usual (for example, about 50 lb/cu ft compared with 56 lb/cu ft); thus self-feeding is improved. Admittedly as a result the silo does not contain as much total silage—although it may well contain more *edible* silage than it would if made by most present methods.

Clamp silos in the field can also be self-fed, with control by a 'ring' electric fence, but unless the ground is well-drained heavy poaching may result without a hardcore or concrete base.

MECHANICAL LOADING AND FEEDING

More attention is now being given to mechanical loading and feeding of clamp and bunker silage. This has the advantage of giving much greater flexibility of layout of silos and yards, and in particular of allowing a settled height of 8 ft or more in the silo. Limitations on intake which may arise from self-feeding are removed and the silage can be mixed with other feed components in a 'complete' ration.

Hand cutting and loading into trailers for feeding has now largely been superseded by the use of tractor grabs and foreloaders, although where long silage is being loaded it still may be necessary to use a knife or mechanical cutter to divide it into blocks. Most foreloaders have the disadvantage that the tines must lift to pull the silage out, and this may let air into the exposed face and cause it to heat and mould. The preferred type has hydraulically-operated curved tines which engage with fixed horizontal tines (photo 41), so that the silage can be pulled cleanly away from the face—although a big tonnage of silage must be moved to justify this more expensive machine. The same applies to fully-automatic silo unloaders (photo 42), but these are likely to be widely adopted as more silage is made, and depreciation may be reduced by using a mobile unloader (photo 43) to serve several feeding units.

These machines can load the silage into ordinary trailers, from which it is unloaded into feeding troughs by hand, or, if one trailer side is removed, by tractor foreloader. But as herd size gets bigger, the main method used is likely to be the side-unloading forage box (photo 44). The main merit of the forage box is that it allows other feeds to be mixed with the silage before feeding; the different feeds are loaded in rough layers one on top of the other, and are then thoroughly mixed as they are pulled forward into the beater arms and thence by the cross conveyor into the feed trough.

Results from this method of feeding on a Lincolnshire farm were noted in Chapter 9. In March 1972 this herd was averaging 5 gallons a day without any parlour feeding. Further, where several different groups of animals are to be fed, a 'complete' ration for each group can be delivered by putting the appropriate amounts of the different feeds into the load to be fed to that group; this is particularly useful where a herd is separated into high, medium and low-yielding groups in different yards—and would be most difficult to apply with self-fed silage.

Nutritionally there may also be real advantage in each animal taking in a complete feed mix all the time, compared with conventional feeding where concentrates are fed twice daily, but this needs further study. Certainly stock find it very difficult to select the different feeds from these moist mixtures based on silage, compared with the easier selection between, say, compound pellets and rolled barley mixed together. It is worth noting that the forage box allows 'surplus' feeds such as cannery waste, stock-feed potatoes, beet tops, etc to be fed much more easily than when each of these must be fed separately.

An interesting alternative is the chain-flail manure spreader, which has been modified at Shuttleworth College and used for feeding silage; this distributes the silage against a wall for feeding, and it should be possible to mix other feeds, such as dried grass and barley, by spreading them on top of the silage before unloading.

SILO SIZE AND CAPACITY

The numbers of stock will determine the physical size and capacity of the silo, but when planning a system it should be recognised that the type of feeding method employed will influence the *shape* and dimensions of the silo. If feeding-out progresses too slowly down the length of the silo, then oxidation on the open face will cause heating and moulding, and a reduction in palatability. Silages vary in their stability when exposed to air, but generally the drier crop with a lower acid content is less stable than is wetter material with a pH below 4·2.

When self-feeding is practised the aim should be for the silage face to be fed back by at least 4 in daily—say a 6 in width for each dairy cow feeding at a 6 ft high face with silage of normal density. When silage is loaded out mechanically progress down the silo

41. Silage can be rapidly removed from a clamp with a 5 ft silage or manure fork used in conjunction with a 3-tined hydraulically-operated grapple fork.

42. NIAE prototype bunker silage unloader. This unit can handle long as well as chopped high moisture-content silage. The "clean" silo face reduces the effect of secondary fermentation.

43. Tractor-mounted clamp silo unloader, which can handle short chopped silage at rates of 15-20 tons an hour.

44. Self-unloading forage boxes, some of which can unload from either side, may be used to feed silage, or a balanced mixture of silage, cereals and dried grass, to stock in the field or in yards.

should be more rapid, of the order of 9-12 in daily, because machines tend to loosen the silage face.

In cases where two silages are to be mixed and fed together, then the dimensions of both silos must be adjusted to suit this feeding procedure if secondary deterioration is to be avoided. For a feeding period of 150 days each silo would be at least 100 ft, and possibly 150 ft, in length, and of the width needed to hold the amount of silage required. Generally speaking, such silo dimensions are much longer (and probably narrower) than those commonly in use. The feeding programme will be much more effectively managed if such calculations are made, and acted upon, when the silo is being planned.

The most fully-mechanised systems of course are those from a tower-silo, with the top- or bottom-unloading mechanism loading into an auger or shuttle conveyor directly to the feeding area. Silage from more than one silo, and other feeds, can also be mixed automatically before feeding.

However, this system can result in some lack of flexibility in building layout, and it is interesting that many operators are now unloading tower silos into forage boxes—giving the same possibilities for 'complete' feeding, and feeding at some distance from the silo, that are being exploited with bunker silos. But this trend, if continued, would remove yet another advantage of the tower over the bunker as a method of storage.

Dried grass

Our opinion must have become clear that, in most cases where grass is dried, this will be as a cash crop on a basically arable farm, rather than on the livestock farm where it is to be fed.

Thus if grass-drying *does* expand, most of the dried grass fed will be purchased as part of the 'concentrate' ration used on livestock farms. It will mainly be delivered in bulk (delivery in bags adds about £2 per ton to the price), and must be stored in a dry, well-ventilated area to reduce the risk of moulding (as far as is known the mould is quite harmless to stock, but we are sure should be avoided). From there it can be taken by hand or by tractor fore-loader to the feeding troughs; feeding in the milking parlour is not advised, as the slower rate of eating of dried grass pellets than of normal concentrates can slow down the milking routine.

FEEDING IN PELLET FORM

It is now recognised that, provided some long roughage is fed, large particle size in dried grass is not needed (p. 142). Most dried grass is therefore likely to be in cobs or pellets of fairly small size

($\frac{1}{2}$-$\frac{3}{4}$ in), which are readily eaten by stock, although lambs and calves may prefer a smaller size ($\frac{3}{8}$-$\frac{1}{2}$ in). As we have noted (p. 145), the feeding value of dried grass is greatly improved when it is fed in combination with barley, and it is in this form that stock farmers are most likely to use it, possibly as a compound pellet, containing dried grass, barley and minerals (as fed in the MLC Bull Tests). But cubing machinery producing this compound works more slowly than with conventional compounds, and this puts up costs; further, only barley from the previous year's harvest can be freely available over the main drying season.

So there is now some interest in the feeding of mixtures of dried grass pellets and mineralised barley pellets. With these the compounder would load into the bulk delivery lorry the required weights of dried grass pellets (ex drier) and mineralised barley pellets, which would be well-mixed at unloading; with pellets of say $\frac{5}{8}$ in there would be little chance of feed selection by the animals fed. Alternatively, the feeder could purchase dried grass directly from the drier, and make the barley pellets with fixed or mobile contract cubing equipment. These compound or mixed pellets can then be fed directly from troughs, or mixed with silage and other components and fed out from a forage box.

Thus conserved forages can be fed in many different ways. But in each case, if they are to be used to best effect so as to reduce the feeding of conventional concentrates, every effort must be made to ensure that they are of high 'quality', and that in the case of hay and silage the amount animals can eat is never restricted. Finally, as livestock units become larger, silage, and to some extent hay, is increasingly likely to be fed as a major part of complete mixed rations, rather than entirely separate from the concentrate part of the ration, as at present.

Chapter 11

CONSERVATION IN DIFFERENT FARMING SYSTEMS

THE MAIN emphasis in this book has been given to barn hay-drying and to ensilage in clamp and bunker silos, because we believe that, these methods offer the greatest prospect of an all-round improvement in forage conservation. But we must be realistic.

Some 85% of the forage conserved in the UK at the present time is in the form of hay, with only 15% as silage and less than 1% as dried grass. Most of this hay is field-made, much of it under conditions which make a low yield and a low-quality product almost inevitable; it would be optimistic to expect a sudden shift to barn-dried hay and silage, and the immediate aim must be for an improvement in field hay-making.

Quite minor changes would help. Over 30 years ago it was shown that tedding the swath *immediately* after mowing greatly speeds up drying rate, yet thousands of acres of smooth swaths can be seen round the shires any day in July; easy adjustment of the drive ratios to reduce the rotor speed can reduce losses in flail haymaking; taking the grass from headlands for silage allows the rest of the field to be cut in lands for haymaking, and avoids delay in baling because the grass cut on headlands is often 'bunched', and shaded from sun and wind, and so dries more slowly than that on the rest of the field.

For most livestock farms, however, haymaking can only be an intermediate step. The quality of field-hay is so much at the mercy of the weather that it is almost impossible to *plan* for it to provide more than maintenance in winter. If conserved forage is to provide genuine production in addition to maintenance, so as to reduce the level of supplementary feeding needed, then a more reliable method must be adopted.

ENSILAGE—THE MAIN CONSERVATION METHOD

We have no doubt that for the great majority of farms the main method should be ensilage, for this alone can cope with storing the huge quantities of grass to be cut at the end of May and in early June. A team of three men, well-organised, can cut and load 80-100 tons of grass daily, with 800 tons of made silage within 10 days a realistic objective; we know of no barn hay-drying installation large enough to deal with 150-200 tons of hay in the same time.

In so far as any conservation method is independent of the weather it is silage, and the possibility it allows of conserving large amounts of grass *to a timetable* is especially valuable. If date of cutting has to be delayed by the weather, not only does the digestibility of the cut grass fall, but the date at which the re-growth is ready for grazing will also be delayed—perhaps critical where stocking-rates are high. Even when the grass has been cut, bad weather can hold up carting from the field if the crop is to be made into hay or wilted for high dry-matter silage, and the shading from the swaths as well as the frequent passes by machines over the field can further reduce the re-growth. The aim must be to cut the crop at the right stage and to remove it from the field as quickly as possible—and in most cases this means silage made without a lot of wilting.

BARN HAY-DRYING OF LATER GROWTHS

Once the main spring flush of grass has been conserved, later growths for cutting are likely to be lighter crops, and some of these are cut in mid-summer, at a time when wilting is more practicable than in the spring. Both these features make the adoption of barn hay-drying of interest for storing these re-growths. The tonnage to be stored will be less than in the spring, so that some type of storage drier, either for bales or for chopped hay, is suitable. Costs will be reduced if the machine which earlier made the silage can be used for harvesting the grass for hay.

The aim of making some barn-dried hay is that this feed should have a particular role in the winter-feeding programme—for rearing young stock for example, for which silage is not ideal. There seems little advantage in making a small amount of hay to feed with the silage fed to the dairy herd; thus where no such special role is evident, it is probably more useful to continue with silage-making. And here a further advantage of the Dorset wedge system appears. The loading slope is quickly and completely sealed at the end of the first bout of silage-making in the spring. If this is done properly, there should be perfect silage directly beneath the plastic sheet when this is pulled back, say in July or August; the second-cut crop then can be loaded in straight away, without first having to remove the 6 in or so of rotted material still found at the surface of many silos. In this way the 'wedge' can be extended at intervals during the season, and this allows quite small quantities of grass to be added to the silo, quantities that are often now wasted because it is so inefficient to make a separate small silo.

These later lots of grass, particularly those cut in the autumn, are seldom of the top quality that is cut in spring; they are therefore well adapted to storing in an outdoor clamp silo (p. 111), and may be conveniently fed to outwintered stock that do not need the best conserved feed. To be effective each new lot of grass must be filled as we have already described, with careful sealing as soon as the last load is put in place, and with the sides at a shallow slope so that the final covering stays on the plastic sheet.

BIG-SCALE UNITS AND TOWER SILOS

With the larger livestock enterprise, and in regions of only moderate rainfall where wilting is more reliable, there will also be wider use of tower silo systems. But to exploit to the fullest extent the mechanisation which towers allow, the scale of operation is important; specialised equipment, including a precision-chop harvester and a dump-box and blower, is needed to fill one tower; with efficient planning to give a spread of cropping, the same machines can equally well fill three or four towers. The greater management skill needed to fill towers than clamps or bunkers also indicates an advantage of scale—though this is not to suggest that skill is not needed for all methods of silage-making!

This shift from hay towards silage-making with the increase in the number of animals kept is clearly shown in Table 20. More than half the dairy farms recorded which kept more than 60 cows used clamp and bunker silage as the predominant methods of conservation, while 11% of herds with over 100 cows were fed from tower silos. The introduction of improved silage techniques, coupled with the continuing trend towards larger herd-size, could thus make silage the main conservation method on dairy farms within a few years.

No similar data are available for beef cattle, but clearly hay is the main conserved forage now fed. With the move towards marketing at a lower age (over 60% of cattle are now slaughtered at under two years of age, compared with only 40% in 1960) it is necessary to feed in winter so as to get daily gains from 1-2 lb per day, in place of the much lower gains of the winter 'store' period. With cheap cereals available it has not yet been necessary to improve the quality of the hay that is fed, in order to get these rates of gain; but as cereals become more expensive the beef feeder, like the dairy farmer, will need to rely more on conserved forage and less on cereals to remain profitable—and to do this he should feed less hay, and more silage.

ALL-YEAR-ROUND SILAGE

Most of the discussion in this book has been with conservation as a method of storing forage grown in summer for feeding in winter. But the possibilities that silage offers for full mechanisation, both from field to silo and from silo to feed-trough, have now led to consideration of systems based on feeding silage throughout the year, on the lines of some North American feed-lots. While some of these units feed mainly grain, others feed mixed rations, including silage made in surface clamps or bunkers which may hold up to 20,000 tons; similar feed-lots, based on maize silage, are already operating in Italy and other European countries.

For several reasons we doubt whether this will be a major development in the UK. First there is the question of scale; it seems unlikely that very large feed-lots will be acceptable in this country, where environmental problems will be of increasing importance, and where the disposal of the 'effluent' from even a few hundred animals already creates problems. Secondly, feed-lots fit into a clearly-stratified system of feeding for fattening the weaner calves from the huge suckler-cow herds of the North American plains (or in the case of Italy, feeding large numbers of calves from Southern Germany, which the local farmers have not yet learned to feed successfully). While many weaned calves and Irish stores *are* fed in this country, they are finished on farms in every lowland county, and mostly sold to local markets—in contrast to the movement of beef cattle towards the populous east coast of the USA. Finally, there is little evidence that beef or dairy cattle can be fed indoors more cheaply in summer than they can be grazed outdoors—and there are also great areas in this country for which grazing is the only effective method of land-use, because they are not suitable for mechanised farming.

Grassland in the more remote areas presents social, rather than agricultural, problems in its continued use for farming, but much—as in Cheshire, Somerset, the Midland counties, etc—has great potential for summer grazing yet with a proportion of the fields level enough to be cut for conservation for winter. If farmers in these areas can improve their overall grazing and conservation management so as to feed animals more profitably than they could be fed in a feed-lot—and we believe many of the techniques to do this are now available—then summer grazing combined with indoor winter feeding will remain the dominant method of livestock production.

THE ROLE OF DRIED GRASS

As we have already noted, it is important to feed as many animals on the farm as buildings and labour will allow in order to increase the 'size of the business'. However, as summer stocking rate is increased, it may be more difficult to ensure early and late grazing in the spring and autumn, so that the length of the 'grazing' season may decrease somewhat. High stocking rates may also reduce the amounts of 'surplus' grass available for cutting, and so effectively limit the amount of silage available for feeding in the winter; this means that other feeds will have to be purchased to maintain high levels of winter production. These are the farms whose economies could be helped by feeding the dried crops (mixed with barley) cut from the grass and other break-crops grown on arable farms. Many arable farmers may prefer to 'cash' their break-crops in this way rather than having to set up their own livestock units.

Table 12 indicated the likely costs of setting up a grass-drying enterprise. These seem enormous sums—until they are considered on a per-acre basis. The $2\frac{1}{2}$ ton/hour drier should produce upwards of 5,000 tons of dried grass annually, the crop from some 1,200 acres, and representing an investment of perhaps £80/acre. This is much less than the investment in any feasible livestock enterprise where, even with beef cattle, well over £100 per acre may be needed for livestock alone. Clearly grass drying as a method of using break-crops must also be considered in terms of the return on the capital invested; but this point is made here because the large area of land which can be served by a single grass-drying unit *would* require a great number of livestock to consume the forage grown, either by grazing or by conservation.

Here as in other parts of this book we have *nearly* entered into discussion of the economics and investment policies of different conservation systems, and of the feeding of conserved forages in place of other feed resources. We considered including a section on economics; but to have done so just at the time of possible entry in the EEC would have been to ensure that any conclusions would be out-of-date before they were printed; even as the manuscript was being completed the fertiliser subsidy was drastically reduced, and the basis of farm investment grants was changed!

Thus we decided to consider only possible economic *trends:* the strong possibility, within the EEC, that feedstuff prices will rise more rapidly than the farmer's returns from the meat and milk he sells, so that he will need to make better use of forages; the certainty that fewer men will be paid higher wages, which will encourage greater mechanisation, and the more intensive use of machines; and the need, therefore, that each unit of forage handled should be of reasonably high value, compared with the low-grade feed so often conserved at present. We hope that the information and ideas brought together in this book will help to ensure this latter objective.

APPENDIX 1

FURTHER READING

Chapter 3   Nutrient allowances and composition of feedstuffs for ruminants. ADAS Advisory paper No. 11.

Chapter 4   Grass species and varieties. Grassland Research Institute, Technical Report No. 8.

NIAB Farmers Leaflets (1971/72) No. 16 Recommended varieties of Grasses, and No. 17 Grass for Conservation.

Chapter 5   Mechanisation Leaflet No. 4. MAFF—Machinery for swath treatment of hay.

Chapter 6   Green-crop drying. Farm electrification handbook No. 15. Electrical Development Association.

Mechanisation Leaflet No. 16 MAFF—Bale handling equipment and systems.

Mechanisation Leaflet No. 23. MAFF—Barn-hay drying.

Chapter 7   Silage. MAFF Bulletin No. 37.

Mechanised handling and feeding of tower silage. MAFF Short Term Leaflet 126.

Mechanisation Leaflet No. 13. MAFF—Forage harvesters.

Mechanised handling and feeding of bunker silage. MAFF Short Term Leaflet 103.

Automatic feeding—Farm electrification handbook No. 9. Electrical Development Association.

APPENDIX 2
METRIC CONVERSION TABLE

All data given in this book are in British Units: below are a few useful conversions to metric units.

Weight	1 ton	=	1016 kilograms
	1 cwt	=	50.8 kilograms
	1 lb	=	0.45 kilograms
Length	1 mile	=	1.61 kilometres
	1 yard	=	0.91 metres
	1 foot	=	0.30 metres
	1 inch	=	2.45 centimetres
Area	1 acre	=	0.404 hectares
Volume	1 gallon	=	4.55 litres
	1 cu ft	=	28.3 litres
Yield	1 lb/acre	=	1.12 kilogram/hectare

APPENDIX 3

Table 1. Moisture Contents at which Hay can be Removed from the Field

Treatment	Mc limits	Swath exposure time, hours
	%	
1. Barn hay drying, using some heated air ..	45–60	8–72
2. Baled, chopped or loose hay dried in barn or tunnel	35–40	24–96
3. (a) Storage conditioning of baled or chopped hay	30–35	48–120
(b) Application of additives, e.g. propionic acid	30–40*	48–120
4. Baling, followed by field conditioning or storage	20–30	48–14 days

*Application at above 35% mc may be influenced more by economic than by technical considerations

Table 2. The Digestibility and Protein Content of Hay made by Different Methods

Date of cutting	Cut grass		Barn-dried hay		Field-made hay	
(1958)	D-value	%CP	D-value	%CP	D-value	%CP
6 June	65.5	11.1	62.5	10.8	58.2*	9.5
17 June	60.5	9.5	52.0*	8.5	45.7*	8.4
25 June	55.2	7.5	52.2*	8.6	41.5*	7.8

*crop affected by rain in the field.

(Data, Shepperson, NIAE)

Table 3. Corrections to The Estimated D-Value of Cut Forage to Allow for Losses in the Conservation Process

Method	Subtract from D-value of forage
Barn-dried hay; good wilting	2
Barn-dried hay; moderate wilting	4
Rapid field hay; good weather	3
Rapid field hay; bad weather	5
Traditional hay; good weather	3
Traditional hay; bad weather	8
Low dry-matter silage; little effluent	0
Low dry-matter silage; some effluent	1
High dry-matter silage; good wilting	2
High dry-matter silage; moderate wilting	3
High-temperature dried grass; direct cut or good wilting	1
High-temperature dried grass; moderate wilting	2

APPENDIX 3 (Continued)

Table 4. Method for Estimating the Starch Equivalent (SE) of Fresh Herbage, Hay, Silage and Dried Grass from The D-Value of The Feed

D-value	Estimated SE content
75	62
70	54
65	47
60	40
55	34
50	28
45	23
40	18

(Data, Alderman, ADAS)

Table 5. Daily Gains by 4-Month-Old Steers Fed Dried Grass Cut on 3 Dates in Spring, and Either Chopped or Pelleted

	12 May		Date of cutting 3 June		28 June	
D-value of grass	73		67		59	
	Chop	Pellet	Chop	Pellet	Chop	Pellet
Intake lb DM/day	8.2	8.8	6.6	8.2	5.7	7.0
Daily gain lb/day	2.2	2.6	1.7	2.1	1.0	1.4

(Data, Tayler & Lonsdale, GRI)

Table 6. The Dry-Matter Content of A Crop After Different Conditioning Treatments

Treatment	Dry Matter % (27 hr wilting)
Cutter-bar mower—turned	24.3
Cutter-bar mower—tedded	26.1
Flail mowing, 5 ft and 6 ft	28.6
48 in flail haymaker	31.6
Cutter-bar mower—crimped	32.3
40 in flail haymaker	35.0

Table 7. Equilibrium Values for Hay Moisture Content and Air Humidity

RH of air %	Hay moisture content %
95	35
90	30
80	21.5
77	20
70	16
60	12.5

APPENDIX 3 (Continued)

Table 8. The Effects of Chopping and Wilting on The Weight of Trailer Loads

Dry matter content of crop	Type of Harvester	10 ft × 6 ft × 6 ft (360 cu ft)		12 ft × 7 ft × 6 ft (504 cu ft)	
		Wet weight (cwt)	Dry weight (cwt)	Wet weight (cwt)	Dry weight (cwt)
Direct cut	—Flail	30	6	42	8
20% DM	—Double-chop	50	10	70	14
Wilted	—Flail	30	9	42	13
30% DM	—Double-chop	45	13	63	19
	—Precision-chop	50	15	70	21

(Data, Shattock & Catt, ADAS)

Table 9. Fresh Crop Requirements for Different Sizes of Grass Driers.

	Evaporative capacity of drier lb water/hr			
	5,000	7,840	10,000	20,000
Wet grass, 80% mc				
Output at 10% mc cwt	12.8	20.0	25.6	51.2
Wet grass required tons/hr	2.9	4.5	5.8	11.6
Wet grass required tons/24 hr	70	108	140	280
Wet grass, 75% mc				
Output at 10% mc cwt	17.2	27.0	34.2	68.8
Wet grass required tons/hr	3.1	4.9	6.2	12.4
Wet grass required tons/24 hr	75	118	149	298

Table 10. The Effect of Crop Moisture Content on Drier Throughout

Months	Crop mc %	Drying load tons water/ ton dried grass	Output of 1 ton/hr* drier tons/hr
April and November	87	5.92	0.59
May, September and October	82–85	4.00–5.00	0.88–0.70
June	80	3.50	1.00
July	75	2.60	1.35
August	78	3.09	1.13

*Evaporative capacity 7,840 lb/hr.

APPENDIX 3 (Continued)

Table 11. Direct Costs per Ton of Dried Grass (1971)

	£
Oil	4
Electricity	1
Labour	1
Repairs	0.5
Interest on working capital 10%	1.5
	£8.0

Table 12. Effect of Scale on the Grass-Drying Operation (1971 costs)

	Drier capacity (rated tons/hr from 78% mc)		
	2½ ton	1¼ ton	12½ cwt
Production per year 2,000 working hours	5,000 ton	2,500 ton	1,250 ton
Acres required at production of—			
4 tons dried/acre	1,250	625	312
5 tons dried/acre	1,000	500	250
Capital cost. Drier, press, field machinery and buildings for drier and crop storage	£110,000	£70,000	£45,000
Depreciation over 10 years— Straight-line write-off per ton per year	£2.20	£2.80	£3.60

Table 13. Milk Production by Friesians (6 Cows and 6 Heifers) Fed Entirely on Silage from 8–15 Weeks of Lactation

	% DM	D-value	Intake of DM (lb)	Average milk yield (gall/day)
Wilted silage	29	71.8	32	2.6
Unwilted silage	20	69.8	24	2.5*

*These cows lost some weight.

(Data, Berkshire Institute of Agriculture and GRI)

Table 14. Approximate Milk Yield from Conserved Forage at 3 Levels of D-value, Supplemented with 4 lb of Barley

Forage D-value	Likely level of intake (lb DM)	SE from forage (lb)	Milk from forage +4 lb barley (gall/day)
75	38	23	7
65	34	16	5
55	30	10	2

APPENDIX 3 (Continued)

Table 15. Daily Milk Production from 8–18 Weeks of Lactation by Friesian Heifers Fed Grass Silage *ad lib*, Plus the Same Controlled Amounts of Either Wafers or Pellets Made from The Same Dried Grass (D-Value, 59)

	Form of dried grass	
	Wafers	Pellets
Ration lb DM/day		
Dried grass	11.9	11.9
Silage	15.6	17.4
Total DM intake	27.5	29.3
Milk yield, gallon/day	3.1	3.7
Butter-fat %	3.7	3.8

(Data, Tayler & Aston, GRI)

Table 16. Milk Production from 12–20 Weeks by Cows Fed a 50 : 50 Mixture of Dried Grass Pellets and Mineralised Barley in Place of a 17% Dairy Compound, as a Supplement to Straw or Silage

Maintenance	Production per gallon	Milk yield gallon/day	% Butterfat
BOXWORTH EHF: dried lucerne 56% D-value			
a 9½ lb straw 6 lb compound	4 lb compound	3.34	3.87
b 9½ lb straw 3 lb barley 3 lb dried grass	2 lb barley 2 lb dried grass	3.47	3.62
BRIDGETTS EHF: dried grass 69% D-value			
a	4 lb compound	3.64	4.02
b cereal silage 15 lb DM	2 lb barley 2 lb dried grass	3.69	3.81

(Data, Strickland, ADAS)

Table 17. The Growth of Young Friesian Steers Fed Maize Silage *ad lib* Plus Supplements of Urea and Dried Lucerne

	Age of animals		
	3–6 months	6–9 months	9–12 months
Diet	Liveweight gain, lb/day		
Maize silage alone	0.7	1.3	2.1
Maize silage +1½% urea	1.2	2.2	2.2
Maize silage +1½% urea +20% dried lucerne	2.0	2.3	2.2

(Data, Wilkinson, Tayler & Thomas, GRI)

APPENDIX 3 (*Continued*)

Table 18. Daily Gains by 15-Month-Old Cattle Fed Grass Silage *ad lib* Plus Restricted Amounts of Dried Grass Pellets or Rolled Barley

	Ration A		Ration B	
Feed intake lb DM/day	Dried grass	7.5	Barley	10.3
	Silage	14.0	Silage	10.3
	Total	21.5		20.6
Liveweight gain lb/day		2.2		2.2

(Data, Tayler, GRI)

Table 19. Daily Gains by Beef Cattle from 3 to 11 Months, Individually Fed on a Rowett-type Barley-Beef Ration, on Dried Grass Pellets, or on Mixtures of Barley and Dried Grass

Ration	Daily feed intake (*lb*)	Daily gain* (*lb*)
Barley beef	18.7	2.37
Dried grass	14.0	1.85
⅓ dried grass/⅔ barley	16.5	2.34
½ dried grass/½ barley	15.9	2.38
⅔ dried grass/⅓ barley	16.4	2.36

*8 animals per group.

(Data; Bee, Drayton EHF)

Table 20. Bulk Feed Storage Related to Dairy Herd Size

	Number of cows			
Method	Less than 20	20–60	60–100	More than 100
	Percent of farms using each method			
Hay	93	70	41	20
Indoor clamp silo	2	15	31	45
Outdoor clamp silo	0	11	20	22
Tower silo	0	2	6	11

INDEX